Forecasting Innovations

Peter Hingley · Marc Nicolas (Editors)

Forecasting Innovations

Methods for Predicting Numbers of Patent Filings

With 24 Figures and 71 Tables

Peter Hingley
Marc Nicolas

European Patent Office
Erhardtstraße 27
80469 Munich, Germany

ISBN 10 3-540-35991-5 Springer Berlin Heidelberg New York
ISBN 13 978-3-540-35991-3 Springer Berlin Heidelberg New York

Cataloging-in-Publication Data
Library of Congress Control Number: 2006931133

This work is subject to copyright. All rights are reserved, whether the whole or part of the material is concerned, specifically the rights of translation, reprinting, reuse of illustrations, recitation, broadcasting, reproduction on microfilm or in any other way, and storage in data banks. Duplication of this publication or parts thereof is permitted only under the provisions of the German Copyright Law of September 9, 1965, in its current version, and permission for use must always be obtained from Physica-Verlag. Violations are liable for prosecution under the German Copyright Law.

Springer-Verlag is a part of Springer Science+Business Media

springer.com

© Springer-Verlag Heidelberg 2006

The use of general descriptive names, registered names, trademarks, etc. in this publication does not imply, even in the absence of a specific statement, that such names are exempt from the relevant protective laws and regulations and therefore free for general use.

Cover: Erich Kirchner, Heidelberg
Production: LE-TEX, Jelonek, Schmidt & Vöckler GbR, Leipzig

SPIN 11786023   Printed on acid-free paper – 42/3100 – 5 4 3 2 1 0

Preface

This book studies patent applications as indicators of innovations. In the time series literature, innovations are unpredictable steps in a process that are not explained by the mechanistic trend that is imposed on data via a deterministic model. Similarly, there is a degree of unpredictability in the development of numbers of patent filings, because they are a proxy for human inventive capacity, with specifications that are by definition a-priori unknowable. Statistical generalisations are required.

Patent offices need to plan for future resource requirements, even though they are governmental or intergovernmental institutions that are not normally set up for the purpose of making profits. The development of demand for patent rights is also of interest to others who seek to map patterns of technological development.

The European Patent Office (EPO) regularly makes forecasts of its future workload in terms of the expected numbers of patent filings as a key indicator. A number of well established and conceptually simple statistical forecasting techniques are regularly applied. The methods are mainly based on straight line trend analyses and surveys. We have both been involved in this work since the early 1990s. While it is disappointing that it is the nature of forecasting that predictions only rarely agree very well with later out-turns, we have always enjoyed the challenge to find ways to improve the methods.

Several years ago, we realised that patents had become popular objects for economic study and that the EPO might have been missing out on better methods that were emerging in the literature or in practice elsewhere. An advisory group of experts was set up to advise on improvements in forecasting methods. The central recommendation of this group was for an external research programme. This book summarises the results of the programme, which ran from 2001 to 2005.

Chapter 1 sets the background to the EPO filings forecasting problem, which is followed by Chapter 2 that explains the research programme and mentions several techniques that are explained and used later in the book. Chapter 3 reviews econometric and time series approaches to forecasting. Five contributions (Chapters 4–8) then describe the research projects. Methodologies and software are discussed as well as models, data and results. Finally, Chapter 9 provides a review, with indications of ways that we hope to put the ideas into action to augment the existing methods.

Aspects of the research will be applicable in other settings. We believe that there is something here for everyone with an interest in business forecasting. By focusing on various aspects of a single theme from different angles, more can be gained than from a text book of unconnected forecasting recipes.

We are lucky to have good contributors. Dietmar Harhoff is professor at the Institute for Innovation Research, Technology Management and Enterprise, Ludwig Maximilians University, Munich, Germany, and helped to draw up the specifications for the research programme. Nigel Meade is professor of quantitative finance at the Tanaka Business School, Imperial College, London, and has been a channel to connect us to the wider forecasting world. Knut Blind and Rainer Frietsch are, respectively, senior researcher and scientist at the Fraunhofer-Institute Systems and Innovation Research Centre in Karlsruhe, which frequently carries out studies on intellectual property and patenting. Gerhard Dikta is professor for mathematics and applied mathematics at the Aachen University of Applied Sciences, and has worked both inside and outside the EPO on forecasting and time series analysis. Walter Park is a professor of economics at American University, Washington D.C., and has a continuing interest in econometric approaches to intellectual property.

We would like to thank Jacques Mairesse, Nigel Meade and Ulrich Schmoch, who constituted the original advisory group of experts. Colleagues at the EPO then helped to take the programme from possibility to reality and on to its successful completion. We particularly thank Steen Andersen, Robert Dunstan, Wolfram Förster, Jacqueline Gnan, Frank Hafner, Wolf Marder, Ciaran McGinley, Bernard Paye, Alain Pompidou, Alexei Sevruk and Elisabeth Smith. We are also grateful to the budget and finance committee of the administrative council of the EPO for sponsoring the research programme.

We respectfully acknowledge the benefit of advice received over several years from the programme participants and from other colleagues both inside and outside the EPO. All opinions that appear in this book are those of the named authors of each contribution, and do not necessarily coincide with the views of the EPO itself.

Munich,

Peter Hingley
Marc Nicolas

April 2006

Contents

Chapter 1   Background 1
P. Hingley and M. Nicolas

Chapter 2   A research programme for improving forecasts of patent filings 9
D. Harhoff

Chapter 3   From theory to time series 27
P. Hingley and W.G. Park

Chapter 4   An assessment of the comparative accuracy of time series forecasts of patent filings: the benefits of disaggregation in space or time 41
N. Meade

Chapter 5   Driving forces of patent applications at the European Patent Office: a sectoral approach 73
K. Blind

Chapter 6   Time series methods to forecast patent filings 95
G. Dikta

Chapter 7   International patenting at the European Patent Office: aggregate, sectoral and family filings 125
W.G. Park

Chapter 8   Micro data for macro effects 159
R. Frietsch

Chapter 9   Improving forecasting methods at the European Patent Office 191
P. Hingley and M. Nicolas

References 247

Name index 255

Index 259

Chapter 1

Background

Peter Hingley and Marc Nicolas

Controlling Office, European Patent Office, Munich, Germany

This book contains reports and a review of several projects that were undertaken during a research programme on the improvement of forecasting methods for patent filings at the European Patent Office (EPO). A typology appears in Chap. 2 of the various areas of study that were identified for the programme.

Figure 1.1 shows the annualised historical counts of patent filings at the EPO. There have been continual increases, except for short resting periods at the beginning of the 1990s and at the beginning of the 2000s. A rapidly increasing trend seems to have established itself since 1996. The fitted line in Fig. 1.1 represents a straight line regression that has been fitted to the total filings data from 1996 to 2005, but excluding years 2000 and 2001 where the development was a little aberrant. From the point of view of business planning, the EPO normally forecasts total patent filings over a six year horizon.

It is not our intention to provide more than a snapshot of current forecasts for future filings at this point of time (early 2006). Forecasts age rapidly. The studies therefore concentrate on the methodology for making forecasts. This should largely remain valid at whatever date you happen to be reading.

We provided most researchers with data on filings up to year 2000, which is why the results that are presented mostly take 2000 as the last actual filings year. Figure 1.1 shows that this was a challenging point of time to make forecasts, since there was then a pause in the upward development of numbers of EPO filings in 2002 and 2003. This was presumably caused by a reduction of investment in innovation after the collapse of the biotechnology/dot-com bubble of the late 1990s. Many of the forecasts presented here consequently over-estimate patent filings for the years beyond 2001. This is no more than can be expected of regression based models that have been fitted over a rapid growth phase. It illustrates the challenging nature of making forecasts, rather than any deficiency in the approaches that have been used.

It might be possible to construct a single large portmanteau model that could take account of the complete set of economic drivers of the patenting process. But this would be a difficult beast to control and could suffer in the presence of any degree of misspecification. According to the principle of parsimony, simplicity is a virtue to be pursued in modelling as long as the fitted data remain consistent with the general patterns of development. In Chap. 9, we will describe how we see modelling fitting in to the forecasting exercises that already underpin the overall EPO budget planning process.

Now, before we start, we would now like to introduce some relevant issues about the patenting process and the statistics that relate to it[1].

In order to apply for the intellectual property protection that is provided by the patent system, applicants may use the following types of procedures, or combinations of them:

- National patent procedures
- Supranational patent procedures, comprising: regional procedures[2] and the Patent Cooperation Treaty (PCT) procedure

The nature of patent rights differs to some extent between countries (Bainbridge 1992). For example, differences in the average numbers of claims included in one application is one reason why there are more patent applications in Japan than in Europe or US. The existence of differences in the scope of applicability of patent rights compromises to some extent the ability to compare patents from different countries.

The patenting concept involves a trade-off between inventor and society, in which the applicant discloses the content of the invention in return for the grant of a monopoly right to exploit that invention for a defined period of time. Disclosure is normally achieved by publication of the contents of the application and also of any resulting granted patent. The original patent application will be termed the first filing and is usually made in the inventor's home country. The applicant(s), who may or may not be the same as the inventor(s), can withdraw at any stage of the procedure. If she/he decides to keep the invention secret after all, the application can be withdrawn before publication, which normally takes place 18 months after the first filing.

---

[1] This is only a short introduction. For further details, please see information that is available from the EPO web pages at http://www.european-patent-office.org.
[2] Regional offices are currently: Eurasian Patent Office (EAPO), African Regional Patent Office (ARIPO), Organisation Africaine de Propriété Intellectuelle (OAPI) and EPO.

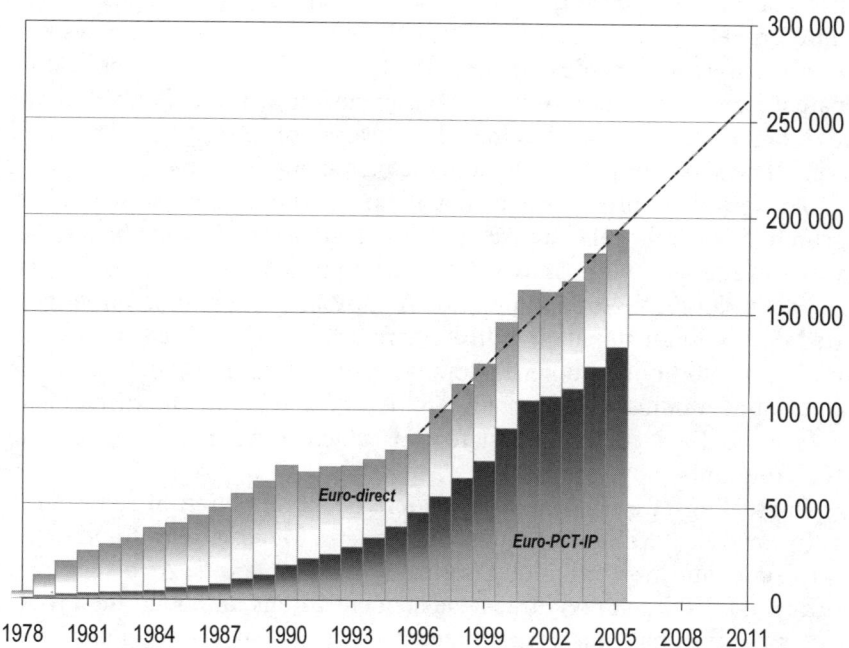

Fig. 1.1. Numbers of patent filings at the EPO per year from 1978 to 2005, with a breakdown by euro-direct and euro-PCT-IP. The trend line is fitted by straight line regression from 1996 to 2005 inclusive, but excluding years 2000 and 2001

For forecasting purposes, it is important to try to get hold of counts of all patent filings and not just those that lead to publications. However, pre-publication information tends to be restricted by patent offices, in order to ensure client confidentiality. Nevertheless each patent office has access to pre-publication data on filings within its own domain. Statistics can be shared between offices without compromising the interests of the clients.

The Paris Convention of 1883 (WIPO 2006a) allows subsequent filings, which claim the priority of the first filing, to be made at other patent offices within the following twelve month period. When patent protection is sought abroad in several countries for the same invention, the applicant is more likely to want to use a supranational procedure, rather than several national systems where he/she would have to replicate actions under several different administrative procedures. The decision regarding subsequent filings in each country is made by the applicant and depends on the expected returns to be made from external markets.

The EPO is a supranational organisation that has examined patents since 1977. It operates according to the terms of the European Patent Convention, or EPC (EPO 2002), and offers a centralised patent granting procedure on behalf of the constituent EPC contracting states. The system currently allows companies or individuals to make a single application for patent rights in 31 countries; and an extension of rights is also possible to five other countries outside the EPC area (as of 1.01.2006). The countries can all be separately designated in the patent application.

The granting procedure is made up of successive phases. Firstly, a search is made for relevant pre-existing documents. This enables the examiner to establish the position of the innovation with regard to the state of the art in its technological field(s). A substantive examination is then carried out to establish patentability in terms of novelty, inventiveness, industrial applicability and other factors. The successful European patent ceases to exist as a unitary whole, but can then become a bundle of national patents in all EPC contracting states for which it remains designated at the time of granting.

First filings are commonly considered as a reflection of efforts in innovative activity, while subsequent filings tend to measure to some extent the internationalisation and globalisation of world markets. As a supranational office, the EPO mainly attracts subsequent filings, although the proportion of first filings has increased in recent years and stands at about 9% of the incoming filings in 2005. It is tempting to extrapolate trends experienced in R&D expenditures and in first filings at various offices to forecast levels of future EPO applications. However, the forecasting problem is complicated by the existence of parallel patent systems in Europe. Applicants can use the national systems, the European system or the international system that has been established by the Patent Cooperation Treaty (PCT). The PCT is a supranational system that is administered by the World Intellectual Property Organisation (WIPO). One PCT application can be made for patent protection in more than 120 countries and 4 supranational organisations, including the EPO for applications designating the EPC area (WIPO 2006b). Before 2004, the applicant had to specifically request a designation for the EPC area, but from 2004 onwards the EPC and all other possibilities for designations in a PCT application are automatically selected. The PCT applications that designate the EPC area are considered as requests for patents to EPO and will be called euro-PCT-IP filings.

So, an application for a European patent can be made either as a direct application to the EPO (euro-direct filing), or as a euro-PCT-IP filing. Figure 1.1 shows the historical development of these two main constituents of EPO filings. Currently, nearly 70% of total filings are Euro-PCT-IP applications. In both procedures, there is an initial stage during which a search is performed[3]. The search report helps the applicant to decide whether he/she wishes to proceed to substantive examination. This typically begins about one year after filing for a euro-direct application, or about one and a half years to two years for a euro-PCT-IP application. These times apply to typical subsequent filings, but are somewhat shorter in cases where the first filing itself is made directly to the EPO as a euro-direct filing. For PCT applications, entry into the substantive examination phase at EPO is termed as entry into the PCT regional phase (euro-PCT-RP) and at participating national patent offices as entry into the PCT national phase.

Our problem is to forecast total filings as the sum of euro-direct and euro-PCT-IP filings. Both types of applications consist of first and subsequent filings. While it is not particularly important to distinguish first and subsequent filings in the use of the forecasted figures, their inherently different characteristics mean that it can be advisable to forecast separately the four combinations of the 2x2 breakdown.

The book concentrates mainly on new suggestions, but the existing methods are reviewed in Chaps. 2 and 9. Chapter 9 also considers how the existing methods can be improved by taking account of the research results. New challenges will be described – for example a breakdown of the EPO planning procedure into technical areas known as joint clusters requires a more detailed approach to forecasting. Also, joint work with other patent offices leads to the possibility that simultaneous modelling of patenting flows across borders might allow for efficient forecasting at each office.

While we do not seek to forecast patent filings at the national offices of the EPC contracting states, there has clearly been a transfer effect over time, away from the national patent examination systems and towards the EPO, where a single application that results in a grant can lead to patent rights in all relevant EPC contracting states. This has to be taken into account in the forecasting process, although the EPO is now sufficiently well established that net transfer these days comes mainly from new contracting states as they join the EPC area.

Figure 1.2 summarises world wide patent filings in terms of the numbers first filings in each major bloc in 2002 and the subsequent filings flows between blocs in 2003, based on reported filings data from various patent offices. The real situation is a somewhat more complicated network of opportunities, since first filings and subsequent filings can take place at most possible pairs of patent offices and supranational systems. Subsequent filings can also be made at the same office that is used for the first filing.

---

[3] In the PCT system, the applicant may also request a preliminary assessment of the likelihood that the invention can finally result in the grant of a patent.

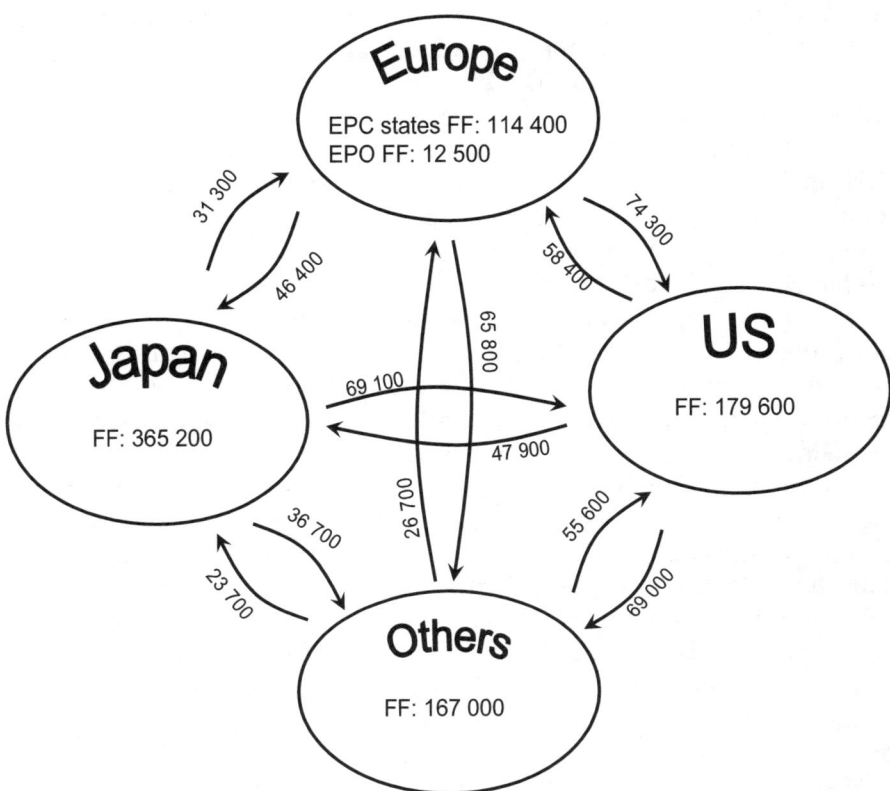

Fig. 1.2. First filings (FF) in 2002 in each main regional bloc: Europe (EPC contracting states), Japan, US and Other countries; and the flows of subsequent filings (SF) between the blocs in 2003. Note that the existence of more than one office in a bloc can lead to multiple counting of patent applications for the same invention. SF flows within each of the four blocs are not shown.

In order to navigate through the maze, it is useful to consider *patent families* data, in which individual priority forming first filings are identified and associated with the resulting world wide filings originating from that priority (see, e.g. EPO, JPO, USPTO 2005; Dernis and Khan 2004, Hingley and Park 2003). Various definitions of patent families are possible, but for us a patent family is the group comprising all patent filings that are made throughout the world and claim the priority of a single first filing, that is itself also included in the group. The set of distinct priority forming filings (that indexes the set of patent families) in principle constitutes a better measure for the set of first filings than the proxy set of aggregated domestic national filings added to first filings at the EPO and WIPO.

Trilateral patent families are a filtered subset of patent families for which there is evidence from published documents of patenting activity somewhere in all three major trilateral blocs: EPC contracting states, Japan and US (e.g. EPO, JPO, USPTO 2005). *Triadic patent families* are another filtered subset that is currently reported in OECD publications (e.g. OECD 2005a). These require an application to the EPO itself, rather than to any patent office in the EPC contracting states, as well as an application to the JPO and the achievement of a grant at the USPTO.

Both the trilateral and the triadic data sets represent only tips of the patent families iceberg. Counts of these underestimate the far greater numbers of first filings that exist throughout the world. This can lead to misconceptions if definitions are not carefully conserved. For example Niiler (2004) gives a table that shows 5 736 patent filings in Germany in 1998. While database sources indicate that there were indeed about 5 300 trilateral patent families arising from first filings from residents of Germany, the same sources indicate a total number of patent families from residents of Germany in 1998 of about 61 400. Regarding reported filing counts by patent offices, German residents made about 46 000 national filings at the German patent office in 1998 and about 20 000 EPO filings (euro-direct + euro-PCT-IP), as well as numerous filings at other patent offices.

We believe that it is not advisable to filter patent family counts when examining patent systems for forecasting purposes. As our main goal is to contribute to the planning of EPO activity, we are particularly interested in all (monolateral) patent families in the EPC area and in bilateral patent families that originate in other blocs and then flow to the EPC area. Within this total, we concentrate on the proportions of the flows that use the EPO rather than other national EPC contracting state patent offices. However, there are unfortunately some difficulties in processing patent families for forecasting, due to lack of timeliness of the recorded data and restrictions of currently available databases to published applications. Therefore we usually apply forecasting models mainly to the reported filings statistics from each individual patent office.

In forecasting patent filings, regression techniques are often applied to historical time series of patent filings. As well as self determining models, such as the straight line that is fitted in Fig. 1.1, it is possible to use prognostic variables that are based on causative factors for patenting that reflect the innovative process. Such variables include Gross Domestic Product (GDP), Research and Development expenditures (R&D) and other economic indicators.

R&D expenditure statistics can be considered at levels ranging from the micro economic level of the firm to the industry and whole country levels. The normal source for collected country based R&D statistics is the Main

Science and Technology Indicators (MSTI) series that is collected and published by OECD (OECD 2005b). Business related R&D (BERD) is often used for patent forecasting purposes, but MSTI also contains series according to several other definitions. Questions arise for analysis as to how to standardise the reported R&D data between currencies and how to discount them over the historical period under consideration in the regression models. It is valuable to collaborate with experts on R&D when attempting to forecast numbers of patent filings using these data. There seem to be at least as many ways to report and standardise R&D statistics as there are for the patent count statistics themselves.

Innovation and patenting take place at different rates in various industries. Classification of inventions is an important task for patent offices in order to support the search phase of the examination. An International Patent Classification system (IPC) has been developed to aid this process (WIPO 2005). Other schemes are also in use to classify company activities in terms of industrial areas, for example NACE (European Commission 2006) and ISIC (UN 2006).

For internal planning purposes, it is also interesting to look at forecasts for particular areas of technology within the whole. The EPO employs technical experts to work on each relevant area of innovation. In this respect the EPO is rather unlike a normal manufacturing environment, where standardisation of products is paramount. There could in fact even be a kind of fractal structure to the scaling pattern of innovations at various organisational levels, which might lead to difficulties in interpretation of forecasting results for patents at the EPO from one scale to any other.

A relevant activity is to watch for and to predict the levels of patent filings in emerging areas of technology. These often correlate closely to usage of the patent system, usually initially while the area is being established, then again after standards are agreed and more players start the development and marketing phase of products. In a sense however, the entire system is in a dynamic equilibrium, with innovations in older areas levelling off and dying as new areas come up. In the world of patents, areas come and go and diversity of topics is the norm. Technology trend watching is a fascinating topic, but it is usually limited to specific areas and it is not our aim to study it here. Rather, we assume a continuing flow of innovative enterprise into whatever areas emerge as ripe for exploitation via the resources of collective human imagination. A system has evolved in society that generates a stream of ideas behind the patent applications, so that in a statistical sense there should be a predictability that can be used in forecasting them. This is the main question that our experts have studied and that is reported on here.

Chapter 2

A research programme for improving forecasts of patent filings

Dietmar Harhoff

Institute for Innovation Research, Technology Management and Entrepreneurship (INNO-tec), Ludwig-Maximilians-University, Munich, Germany

1 Introduction

This chapter describes the starting point of the research efforts described in the subsequent chapters. It draws to a certain extent on an earlier internal EPO report by a group of experts on forecasting.

Filing estimates are of crucial importance for the capacity planning decisions faced by the Office. EPO search and examination capacities are a major determinant of pendencies. Any improvement in the precision of filing predictions will therefore mean a better tradeoff between office efficiency and decision-making lags. In 1999, the EPO asked the external expert group (referred to below as the advisory group) to review its methods and procedures used for predicting the numbers of filings.

The advisory group came to a positive general assessment of the methods and procedures used at EPO to estimate and predict patent filings. The group also made a number of suggestions which were summarized in a report to the Office. In addition to making detailed suggestions as to how the current practice of estimating filings could be improved, the advisory group suggested to set up a continuous research programme. Subsequently, the Office concluded that such a research programme should involve outside experts contributing to an improvement of existing methods. The EPO then asked this author to submit a report in which the structure of a research programme would be outlined.

The report was concluded at the beginning of 2000 and presented suggestions – made on the basis of information available at that time – as to how such a research programme could be set up and which research packages could be specified. The report described five research modules which were to be either delegated to external researchers or undertaken (at least in parts) by EPO experts.

The following sections summarize the suggestions made prior to the inception of the research projects described in subsequent chapters of this book. In Sect. 2, the main motivation for the research projects is described. In Sect. 3, potential approaches towards estimating patent filings are summarized. Sect. 4 describes the proposed research modules, and Sect. 5 discusses the need for coordination among the different modules. Sect. 6 concludes with additional suggestions.

2 Motivation and task description

The basic structure underlying the inflow of filings at the EPO is described in Fig. 2.1. Stages with EPO involvement are highlighted.

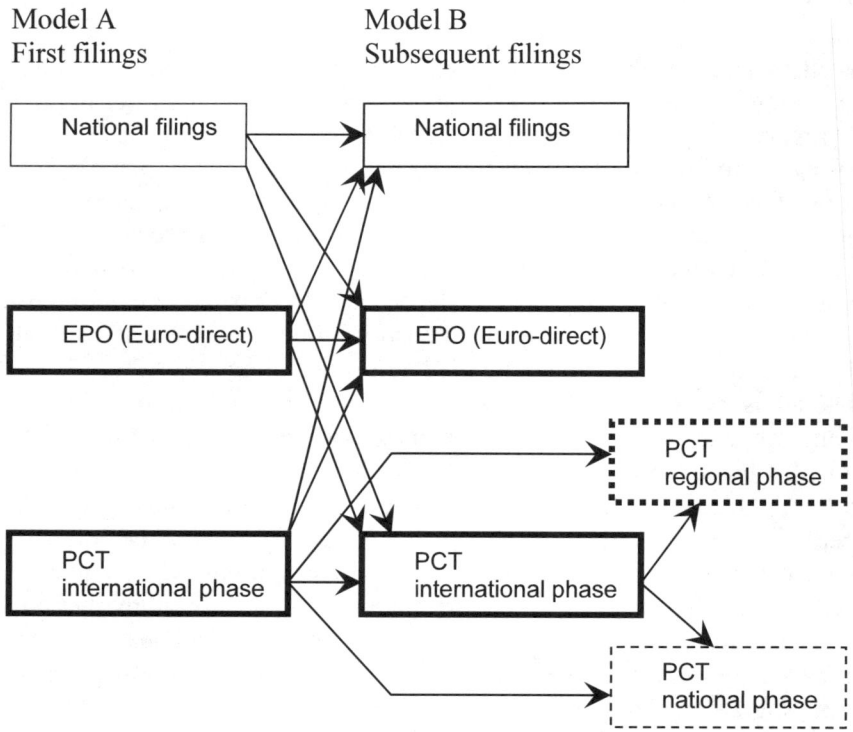

Fig. 2.1. Sources and destinations of filings. The stages where EPO can be involved are highlighted

Existing models for predicting the numbers of filings fall into two categories. The first type (Model A) focuses on first filings that are directly filed at the EPO (euro-direct), at national offices (national filings or PCT-IP filings) or at other supranational offices. The second type (Model B) focuses on subsequent filings, that quote the priority of an earlier first filing and that can also be made at the EPO, national offices or at other supranational offices. Subsequent filings may either enter (or remain in) the euro-direct group, or they may be euro-PCT-IP filings. It is important to realize that (as of 2004), about 86% of all applications reaching the EPO refer back to a national priority, and that the lion's share of these priorities (about 89%) are from Japan, the US and EPC countries.

3 Approaches to forecasting patent filings

3.1 Approaches pursued by EPO and discussed in the report of the advisory group of experts

3.1.1 Survey approaches

The most direct way of eliciting information about future filings is to ask potential applicants about their intentions to file applications at the EPO. The Office has conducted such a survey annually since 1996 and the approach is generally thought to generate highly valuable information. The advisory group views this method as the most promising candidate for short-term predictions, since it does not incorporate past trends into future predictions.[1] The Office surveys two groups of applicants – firstly, applicants that typically file a large number of patent applications per year, and secondly, a random sample of applicants with a relatively small number of filings per applicant.[2]

---

[1] Strictly speaking, this may not be correct if the applicants themselves use methods for predicting future filings which depend on long-run trends. In any case, the respondents of the survey can safely be assumed to have better information on such trends at the firm level than the Office.

[2] Another group of interest are firms which have not been actively patenting in the past, but may be willing to initiate their first patent filings in the future. In principle, the existing data at the Office would permit an analysis of how large this group is in relation to the two other ones.

Patent filings are heavily concentrated – the largest applicants account for a very large portion of total filings.³ Surveying a very small number of applicants can therefore generate information on the development of a large share of total future filings. The Office has also asked the panel firms for more differentiated information on the number of Euro-direct, PCT and total filings. This information may be of increasing importance, since the growth rates for these types of applications differ considerably. Yet, the more complex questionnaire may also lead to a reduction in the number of applicants responding to the survey.

The advisory group has supported the survey approach, and it has recommended to undertake the following steps towards improvements:

- Increase the sample size.
- Simplify and/or redesign the questionnaire in some parts, e.g., in order to obtain explanations from the respondents for the anticipated changes in filings.
- Broaden the information base via an inclusion of an "EPO filing question" in other surveys, such as the Community Innovation Survey.
- Improving the estimation techniques used to predict filings based on the survey results.
- Working towards an increased exchange of information and cooperation between the EPO, the JPO and the USPTO.⁴

Some of these comments will be addressed below, as the improvement of the survey method will constitute Module A of the proposed research programme.

3.1.2 Extrapolation

The extrapolation models involve – in some way or another – the modelling of the time series behavior of patent filings. This may involve a simple smoothing technique (such as exponential smoothing), a regression of filings against time, a univariate time series model (such as an Auto Regressive Integrated Moving Average, or ARIMA model), or a multivariate time series model of filings. The univariate models typically consider the total number of filings at the level of the Office. In other research efforts, EPO

---

[3] Given that patent filings are highly correlated with firm size (which is roughly distributed according to a log-normal distribution), this result is not surprising.

[4] The Office is already engaged in an extensive exchange of information on a number of forecasting techniques. A trilateral global applicant panel survey would also be a good idea.

statisticians have used national R&D data to model domestic filings as a function of past R&D expenditures. In current versions of the applicant panel survey, the Office asks firms about their R&D activities. Again, this provides valuable data for forecasting future filings.

The advisory group encourages the EPO to consider ARIMA models as a valuable complement to the theoretically less complex regressions of filings on time trends. In particular, the external experts recommend to employ extended ARIMA approaches which allow other variables, such as R&D, to be included in the estimation. In principle, it is also possible to shift to monthly or quarterly time series if these contain interesting information on the future development of filings. Such an approach is helpful in that it enables the office to recognize relatively early if the predicted filings differ substantially from the actual ones. These issues will again be addressed below, since this report suggests to dedicate one research module (Module E) to ARIMA, Vector AutoRegression (VAR) and other time series approaches, using aggregate data on filings.

3.1.3 Transfer methods

The third generic type of approaches pursued by the Office uses data on first filings in the national offices and on past subsequent filings at the EPO for the computation of transfer coefficients. This approach is attractive, since the national offices can in principle deliver detailed data on their current filings which - with some probability given by the transfer coefficient - may reach the EPO later, typically at the end of the priority year. The transfer method also lends itself nicely to the modelling of the application path, since separate transfer coefficients can be computed for eurodirect and euro-PCT-IP applications which reach the Office via the national offices and WIPO. Therefore, this report suggests to intensify research towards the improvement of the transfer method as currently pursued by the Office. This work will constitute module D (see Sects. 3.2.2. and 4.5 for more details).

3.2 Additional approaches suggested by the advisory group

3.2.1 Econometric modelling

The advisory group recommends to embark on a number of additional research efforts. These are not meant to introduce drastically new methods, but to refine the existing ones based on new empirical evidence regarding the predictive power of the models. One of these approaches recommended by the advisory group is the modelling of the relationship between patent filings and a broader set of economic and behavioral determinants. The patent literature contains a large number of contributions in which patenting activity (either filings or grants) are modeled as a function of broad sets of variables. These models have been specified at the level of enterprises, industries, and countries. This report suggests in the following section to design two research modules (Modules B and C) that deal explicitly with these types of models. One module is supposed to study these relationships at the firm level, another module is suggested for work on the determinants of filings at the industry or national level.

It is noteworthy that there is a direct relationship between models of this kind at the firm level and the survey information considered in Sect. 3.1.1. R&D expenditures play a particularly important role in this context, since the relationship between the two variables is very strong. Given the time lag between R&D spending and the filing of patent applications, information on future R&D could effectively lengthen the time horizon for which predictions can be made with reasonable precision. The Office has already developed plans to ask respondents in the applicant panel for information about their R&D activities. These data may be an effective complement for the research projects sketched out in this report. R&D data may also be or become available from additional sources, such as the Community Innovation Surveys.

3.2.2 Econometric modelling of filings flows between patent offices[5]

Another recommendation of the advisory group concerns more econometric approaches which distinguish between different types of filings or application paths. These approaches can be seen as natural extensions of the transfer method already practiced by the office. The particular appeal of this recommendation is that it would enable a more refined prediction to emerge from the estimation – in particular one that takes heterogeneity across filings into account. Given that the workloads resulting from various types of filings differ, the microstructure in such predictions would be of particular relevance to the Office. More details on this approach and its implications for the research to be undertaken follow in Sect. 4.5.

---

[5] The report of the advisory group uses the term *structural models*, since these approaches consider the structure of the various patent offices and the flows of filings between them in greater detail than time-series or other approaches. This terminology is avoided here, since the term may be misunderstood. In econometrics, the term structural model is typically reserved for a model in which the regression parameters directly reflect corresponding parameters in a theoretical model. See Chap. 3.

3.2.3 Global models and cooperation with other offices

The advisory group also recommends stronger cooperation with the JPO and USPTO in the development of models which focus on flows of filings between the three offices. In terms of the classification developed by the advisory group, this would involve a more structured approach to transfer models – executed in a cooperative effort between the offices.

Taking this suggestion into account in the design of a research project initiated by the EPO alone would obviously run into a number of problems. The degree to which cooperation between large national and supranational institutions can be realized cannot be affected by the EPO alone, nor can it be planned in the context of this report. This particular suggestion of the advisory group is therefore not taken into account here. However, some of the projects described below may very well profit from co-operation with the USPTO and the JPO. As they are described here, they can also be initiated on a stand-alone basis.

3.2.4 Other recommendations

The advisory group is skeptical about additional approaches such as neural networks or applications of chaos theory. That skepticism is shared in this report. These drastically different approaches will therefore not be considered here. But research work based on these concepts can possibly constitute an entry in the research competition outlined below. In that regard, the research competition offers a degree of openness with respect to drastically new approaches that could not be replicated in a contract research setting.

4 Proposed research projects

4.1 Structure of the research programme

An overview of the projects to be undertaken is presented in Fig. 2 which lists some of the most important tasks within the modules, and (in the bottom field of each rectangle) the party or parties to be engaged in the particular project.

This section lists details of the research to be undertaken in each of the five modules A, B, C, D, and E. Furthermore, it briefly summarizes the research questions to be tackled, the statistical and econometric methods that may be applicable, and the types of data needed for a completion of the analysis. The segmentation of the overall research work in five modules follows a number of simple design principles. Firstly, the modules should be reasonably balanced, i.e., of roughly equal weight. Secondly, while some coordination across modules is required and potentially very productive, the modules can be delegated to teams which work separately on their respective task. This is not to say that the tasks are completely independent – each of the modules can profit from some coordination and coupling with the other modules. As will become clear, some of them have particularly strong links, either because they will have to employ the same or very similar data sources, or because they are likely to employ very similar econometric methods.

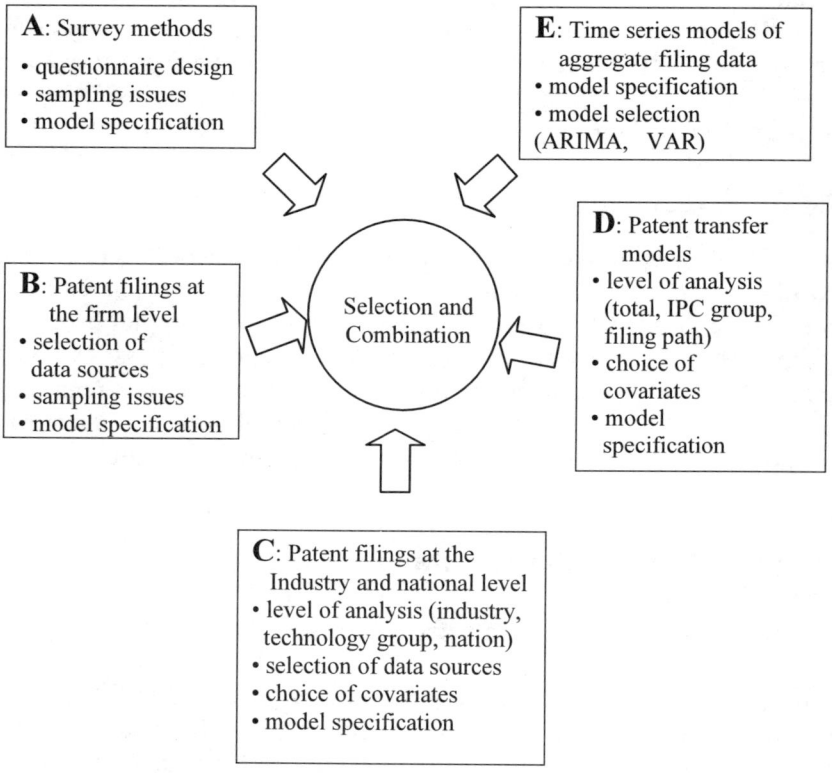

Fig. 2.2. Research modules

The coexistence of various methods which generate different predictions regarding the number of filings in future years also requires a systematic method of selecting or combining these estimates. In principle, this task could be delegated to an additional research module. Given that the coordination requirements for this module would be exceedingly high, this task should either be tackled once the research results have been generated; even better, this task should be considered an activity that is directed and undertaken internally by EPO experts who are in contact with the research groups.

In three out of the five modules (B, C and E), this report recommends that the research work be delegated to external teams which are then monitored by EPO experts. In one case (Module A), the research is likely to be conducted by Office experts with some limited help from outside sources. Finally, in one other case (Module D), the research undertaken will have to be shared about equally between the external and internal researchers. In any case, monitoring and coordination by EPO experts will prove to be critical, in particular when the teams are supposed to take trilateral inputs into account. This is most likely to be needed for Modules A, C, and D.

4.2 Module A – Survey methods

The most direct method of obtaining an estimate of future patent filings is to survey applicants. This method is particularly appealing in the light of the strong concentration of filings within a group of relatively large corporations. Therefore, it is paramount to have precise estimates of the number of applications coming from large applicants. Moreover, by surveying a representative sample of applicants with relatively small numbers of filings, it becomes possible to generate predictions for that group as well. Survey data are particularly valuable when the structural relationships that have guided patenting behavior in the past are subject to major change. While much of the efficacy of the survey approach relies on an efficient execution of the respective survey, a number of interesting scientific research questions remain, e.g.:

- How should survey questionnaires be designed in order to elicit a maximum of information and – at the same time – a high response rate.
- Which firms should be included in the samples, and how large should the samples be given the resource cost of increasing the sample size.
- How should the estimates obtained in the survey be combined to generate an estimate of the overall number of filings expected in some future year.

Leaving aside the third question, which is a classical statistical problem, the first two questions require in-depth knowledge of the previous surveys undertaken by the office. This report therefore suggests to conduct the research tasks of this module within the EPO. External support is feasible and possibly constructive in the following areas:

1. The optimal design of the questionnaire should match the typical information structure of the respondents. Data that are not easily available to the respondents are unlikely to be generated through search and analysis. Hence, to the extent that this step has not been undertaken already, it would be productive to interview a number of survey participants with respect to the data questions they can answer easily and with great precision. This information could guide the EPO in its difficult trade-off between comprehensiveness of the questionnaire and frequency of missing items or even reduced response rates.
2. The form in which the survey is administered could conceivably be subject of research. Internet-based questionnaires have been shown to be quite effective and popular among some respondents. They also offer cost advantages and allow online error tracking.
3. It may also be interesting to consider which incentives can be provided to the respondents of the survey in order to encourage participation. These could involve, for example, extensive reports about future trends in patenting, or free issues of products (e.g., CD-ROM data) that the Office makes available for a fee to the public at large.

The advisory group has recommended that research and work leading to improvement in the survey method should be conducted mainly in-house. This report concurs, given that the coordination of this particular task is relatively burdensome. However, for questions and support in the area of survey design, the Office may want to enlist support in the form of thesis work to be undertaken by graduate or undergraduate students with prior training in empirical methods.

As was mentioned before, the task of generating predictions and measures of dispersion for the expected number of filings is a classical statistical problem. However, the problem may not be trivial, depending on the particular structure of the survey. Again, this report recommends to set some resources aside in order to enlist help from statistical and econometric experts, or – given financial constraints – from graduate and undergraduate students undertaking supervised thesis work in this area. For example, the use of bounded influence estimators as recommended by the advisory group can become the topic of a thesis in statistics or econometrics. A particular expertise of the patent system is not required in these applications, hence, the group of possibly interested researchers is considerably larger than it is for the following modules.

Steps towards a survey conducted jointly or cooperatively by the EPO, the USPTO, and the JPO are currently not included in the concept for this research module as they may depend on factors that cannot be assessed in detail in this report.

4.3 Module B – Patent filings at the firm level

This module is intended to generate research results on the impact of economic and behavioral determinants on patent filings at the firm level. The motivation underlying this research is the same as for Module A. Given that a small number of applicants at the EPO (and at other patent offices) account for a large fraction of total filings, a model describing the behavior of these firms might be a powerful instrument in predicting future patent filings. Moreover, for these large applicants it is typically possible to obtain accounting and other data that may be used in the specification of micro-econometric models of patent filings. A necessary condition for such models to be helpful is to find a robust and statistically well-founded model. In other words, research in this module needs to identify

- The set of covariates to be included in the econometric models.
- Data sources containing the covariates determining filing at the firm level.
- Means of updating these data sources such as to allow EPO to use the methods developed here in regular intervals.
- The econometric estimation procedures best suited for predicting filings at the firm level.

Data collection is the most critical issue for research within this module. It requires researchers to make difficult tradeoffs between data availability and maintenance on the one hand, and methodological and theoretical considerations on the other. Ideally, the data used here should come from standard sources which will reliably produce updates of the data. A restriction to applicants with headquarters in Japan, Germany, France, UK and the US would conceivably lower the complexity of the data collection process without introducing large biases.

The data to be collected for these firms should ideally include information on R&D expenditures, value-added, sales, the number of employees, and other standard accounting information. The most crucial variable is clearly R&D. While R&D expenditures have to be reported by publicly traded US corporations to the Stock Exchange Commission (SEC), they are not part of standard reporting in the UK, France and Germany. However, various researchers have put together relatively large panel datasets that do include information on R&D expenditures which could in principle be useful for the projects described here.

The business press also compiles data on the R&D activities of the largest international corporations. For example, Business Week publishes an annual report on R&D undertaken by U.S. firms. It will be an important task in this module to identify such data, to arrange for access to them, and to conduct the econometric analyses. This work is obviously related to some of the issues tackled in Module A. If it turns out that the EPO survey will regularly collect data on R&D expenditures and expected changes in R&D spending levels, then modules A and B should be closely coordinated.

4.4 Module C – Patent filings at the industry and national level

The research questions relevant for module C are similar to those of module B. At first glance, it may seem puzzling that such an analysis at the industry or national level should be undertaken. However, it is not necessarily the case that predictions based on a firm-level model will outperform a corresponding model estimated at the level of industries or nations. One argument that may lead to a preference for the more aggregated level is the existence of externalities at the firm level. If the patenting behavior of firms is not only a function of own R&D, but also of the R&D of competitors, then a model at the industry level could conceivably be a more precise instrument than estimates at the firm level are. Moreover, more aggregated estimates also circumvent the need for aggregation of the firm level results.

Models at the industry and national level face a number of challenges, though. Firstly, there is the problem of associating patent applications with industries. The natural classification of patents via their IPC codes does not translate easily into some industry classification. Transition matrices to accomplish this transformation have been developed for some nations (e.g., Canada) but not for a large number of European countries. Thus, mapping IPC classes into industries (e.g., classified by NACE codes) is largely terra incognita at this point. The advisory group discusses the possibility of aggregating various industries to a relatively small number of five or six major groups. This report concurs, since this procedure is likely to reduce complexity and allow for a better concordance between patent and industrial classification. The proper group classifications will have to be determined by the researchers, if possible in accordance to the results from other patent offices in the trilateral group. Industry-specific data, e.g. on R&D, value-added, investment etc. are available, for example from the OECD in its MSTI ANBERD database. Once the industry-IPC classification is determined, the aggregation of these variables should not constitute a problem.

A structurally similar problem emerges for models of filings at the national level. This problem arises in particular if one wishes to model first filings at the national offices, since applicants in smaller EPC countries may prefer to send their first application to one of the large national offices in Europe. The advisory group has recommended to circumvent some of these problems by aggregating a number of relatively small countries to one bloc, and by interpreting this bloc as one observation together with other observations reflecting the larger countries.

Given that the purpose of this research is to model cross-sectional and longitudinal variations of filings, a relatively large set of determinants can conceivably be of value in estimating the models. Typically, the data used in this module will consist of panel data which may – at least for some variables – be considerably "shorter" than the data used in the ARIMA and time series estimates at the national level. It is therefore believed that the approach to be studied here has more similarity with the research work conducted in module B, than with typical time series approaches as they will be discussed in Sect. 4.6 which describes a module (module E) which can be dedicated to "pure" time series estimates.

The research questions to be tackled in module C can be summarized as follows:

- Determine covariates to be included in the econometric models at the industry or national level.
- Identify data sources containing the covariates determining filing at the firm level.
- Study means of updating these data sources in order to allow EPO to use the methods developed here in regular intervals.
- Find a procedure to aggregate patent filings to industry-related groups of patents, e.g. concordance matrices between IPC and NACE codings.
- Identify the model specifications and econometric estimation procedures best suited for predicting filings at the industry and national level.

4.5 Module D – Patent transfer models

Models which explicitly consider the flows of filings between patent offices have been described and recommended by the advisory group. These models can be seen as the natural extension of the transfer models already used in the office, and they are therefore referred to as "patent transfer models" in this report. The existing transfer models used by the Office typically try to use simple, but effective extrapolation techniques which are applied to the transfer coefficients, i.e., measures of the likelihood that a patent first filed in some national office will reach the EPO within the priority year either as a Euro-direct filing or as a PCT filing.

In essence, the patent transfer approach amounts to taking the complexity and heterogeneity much more seriously than is done in the other approaches. Modelling various types of filings explicitly increases the value of information coming from precise predictions, but arguably, obtaining precise predictions becomes a more complex job as well. But since PCT applications are becoming increasingly popular (though not equally quickly for the large applicant nations), the approach takes one important source of variation for workload accumulating in the Office into account. Hence, the challenges to be addressed in this module are to

- Identify a sufficiently detailed, but not too complex abstraction of real workflows at the EPO.
- Explore the tradeoffs involved in modelling euro-direct filings and euro-PCT-IP filings either separately or together.
- Obtain information on data availability, in particular of national filings which become the input data for this approach.
- Specify models capturing important determinants of the change in transition coefficients.
- Estimate these models and select the most appropriate variant for future use in the EPO.

The research work in this module would also bring together the work in Modules C and E with the transfer coefficients estimated here, at least to the extent that these modules generate predictions of national first filings. An important step in the combination of these results has been pointed out by the advisory group – once stochastic transfer coefficients and stochastic estimates of future filings are combined, computing the confidence intervals for the resulting estimates is no longer trivial. Given the importance of having precise confidence bounds, this question will also be considered in this module.

Finally, it seems appropriate to point out that the contextual knowledge required for the research work in module D is considerable. This is particularly true if the module also encompasses attempts to model the secular shift from national filings towards filings at the EPO. Such a historical analysis would be highly valuable, but requires in-depth knowledge of applicant behavior and institutional details of the patent system. This report therefore recommends to treat this project as one in which EPO experts and external experts cooperate particularly closely. This module cannot be delegated to an external team that is not in close and immediate contact with EPO experts.

4.6 Module E – Time series models of aggregate filing data

The final proposed module targets the classical time-series estimation techniques which are already being employed in the Office on a regular basis. The advisory group's report suggests enrichment of the methods used so far by employing additional related economic time series that can shed more light on the longitudinal variation of filings. This is obviously appealing if one takes the lessons learned by the USPTO into account: going from a pure univariate approach to one in which other economic variables were used (in particular R&D expenditures) reduced the forecasting error from about 20 percent in univariate models to about 4 percent in the extended models.

Research issues to be addressed in this module E encompass

- Identification of a set of possible regressors to be used in the time series estimates.
- Specification of suitable models (e.g. Vector AutoRegression, ARIMA, error correction).
- Estimation of the models and determination of the most suitable ones.

The data needs for this particular module are considerably less pronounced than for the other models. Moreover, since the focus will be on prediction rather than causal analysis of the determinants of patent filings, the team working on this module will not require prior knowledge of the patent system.

5 Data needs and internal coordination between modules

EPO can substantially facilitate the projects described in the previous section by supporting the prospective researchers with respect to their data needs. This section lists the data necessary for tackling the research modules. It is recommended that the EPO provide most of the data necessary to undertake the research. This support will enable the cooperating researchers to focus fully on their analytical task, rather than lose time in the search for the required data. Given the restricted financial resources available for the overall project, this support will be a crucial element in the overall design. Moreover, by having some control over the data collection process, the Office can ensure that the econometric models to be developed are not contingent on idiosyncratic datasets which may be hard to replicate in the Office.

In some cases, however, the data collection effort should be delegated to the external researchers. This is particularly relevant for Module B in which the identification and acquisition of reliable sources of firm-level data is a crucial part of the overall module.

The following table summarizes the data needs for each of the modules. In order to facilitate communication between the research groups, it is recommended to produce a database that can be used jointly by all teams. If at all possible, the teams should be allowed to use the data at their home locations.

The table also briefly summarizes which projects need to be coordinated closely in order to avoid duplication of effort. For example, Modules C and D share the same data, but Module D requires transfer data in addition to the data used in Module C. The methods used in Module E have some bearing and relevance for Modules C and D. Modules A and B can gain considerable synergies by working with the same or similar sampling strategies.

Table 2.1. Required data for the research modules

Module	Data required	Module to be closely coordinated with
A	no data required – external team contributes by: i) testing pilot questionnaires ii) exploring data availability in direct consultation with respondents iii) exploring alternative modes of surveying (written, internet, phone)	B
B	patent filings by applicant (same sample as used in Module A) R&D at the firm level	A, C
C	national filings industry level economic data (OECD and statistical offices) R&D spending by industry (ANBERD)	B, D (some links with E)
D	as in module C, plus national first filings and structural data (euro-direct vs. euro-PCT-IP)	C (some links with E)
E	national and EPO filings (monthly or quarterly, annually)	(some links with C and D)

6 Further recommendations

The above modules represent a targeted approach to the improvement of forecasting methods at the Office. They can best be carried out as contract research undertaken by interested and capable scientists. It may be possible to relax some of the financing constraints by explicitly earmarking some of the research work as eligible for supervised graduate thesis work. By specifying the needs of the EPO explicitly in the research contracts, the research work can be focused and highly effective. This is exactly the purpose of the modular approach described above. Necessarily, the research work also becomes very specific in the sense that alternative approaches not foreseen by either the advisory group or by this report may not be pursued. Some remedies should be in place to allow such approaches to emerge. Two possible approaches are listed here – a research competition for young researchers and a regular research conference. Other approaches may be attractive, too, but they would require higher funding levels. For example, the Deutsche Bundesbank has reserved funds for a "researcher in residence." This position is meant to attract eminent researchers in the field of public and corporate finance, and it has apparently proven to be quite successful so far. The EPO could follow this example in order to make research on the economics of the European patent system more attractive.

6.1 Research competition

In order to achieve this objective, an additional suggestion is to introduce a best paper award for research on the determinants of patent filings. This call should be restricted to junior researchers who are – at the time of the call – younger than 35 years of age. This would invoke a kind of tournament structure with prizes to be allotted to the winners of the contest, rather than to all participants who have undertaken the effort. The EPO should provide the basic data for the tournament, e.g., on a CD-ROM, but participants should be free to choose or generate additional data that they think is valuable. This mechanism would set in motion the forces of competition and innovation which may prove to be quite effective from the Office's point of view. The announcement of the winners could take place during a research conference which focuses on issues of patenting.

6.2 Research conference

Probably the least expensive research programme is one in which the scientific community is made aware of research questions relevant to the Office and incorporates some of the questions in the normal course of scientific inquiry. It is clear that scientists will be driven by their own perception of research problems and opportunities. However, it is unlikely that this perception will completely exclude applied problems at the EPO from consideration. In order to allow for such a contribution, researchers need to be alerted to questions of particular relevance. Moreover, the EPO should provide the data necessary to carry out ambitious and path-breaking research in the field of patenting. The EPO should therefore consider the costs and benefits of organizing a conference or regular sequence of conferences in which the participants of the research programme, the participants of the research competition and other scientists play a role.

Chapter 3

From theory to time series[*]

Peter Hingley[1] and Walter G Park[2],

[1] Controlling Office, European Patent Office, Munich, Germany
[2] Department of Economics, American University, Washington, DC, US

1 Introduction

In order to develop forecasting models for numbers of patent filings, it is important to consider that patents protect more than just intellectual property; since they are an intrinsic component of the larger economic picture. This occurs through the process of innovation, technological and scientific change, economic productivity and growth. In this chapter, we will show one particular theoretical formulation of the underlying process that leads to patenting. Then we will briefly review the various types of time series based regression analysis that are available and show how they could be used to fit a model within a framework of econometric methodology that uses techniques of cointegration and error correction. Various approaches to time series could be used and this book necessarily concentrates on only some of them.

2 A theoretical model

This section develops a conceptual model of international patenting flows as a basis for the empirical analysis. The following model is adapted from Eaton and Kortum (1996), Kortum and Lerner (1998), and Park (2001). In this approach, a decision-theoretic model of patenting is formulated where inventors weigh the costs and benefits of filing for a patent. Other, more macro level, approaches are possible; for example, Abdih and Joutz (2005) use time-series methods (including cointegration analyses, see Sect. 3.1) to

---

[*] Frederick Joutz and Robert Trost contributed to Sect. 3.1 during the preliminary phase of the research programme.

derive a knowledge production function which relates patent filings to the past stock of filings and to the number of R&D scientists and engineers.

First some definitions and notation are in order. Let the source country be the country of origin of patent applications (or the country to which patents are granted). Let the destination country be the country in which a patent application is filed (or the country which grants a patent right). For now, we are primarily interested in the special case where the destination is not a specific country but a regional office (such as the EPO).

The variations in international patenting depend on three kinds of heterogeneity: (1) market heterogeneities (some destinations are more attractive than others); (2) invention heterogeneities (some inventions are more valuable than others); (3) heterogeneity between source countries (some source countries are more inventive than others).

Let the source country be indexed by i and the destination country by j, such that $i = j$ refers to domestic patent applications, and $i \neq j$ to foreign patent applications. Let P_{ij} denote the number of patent applications from country i to country j. Each source country produces each period a flow of inventions. Let α_i denote this flow of ideas, some of which may be patentable. Of these, some fraction, f_{ij}, is applied for a patent in country j[1]. Hence the number of patent applications from country i in country j is:

$$P_{ij} = \alpha_i f_{ij} \qquad (1)$$

We will model each of the three components on the right hand of this equation. First, we assume inventive output to be produced according to a linearly homogeneous production function.

$$\alpha_i = \alpha \, R_i^{\beta_1} \, S_i^{\beta_2} \, L_i^{1-\beta_1-\beta_2}, \qquad (2)$$

where $\alpha > 0$ and R denotes the stock of research and development (R&D) capital, S the supply of scientists and engineers, and L labor. The exponents ($\beta_1, \beta_2, \ldots$) represent elasticities: the percentage change in inventive output for a 1% change in input.

It will be shown that those inventions which cross a particular quality threshold will be the ones for which patents are sought in the destination market. To develop this idea, assume that each of the various inventions of the source country can be indexed by its "quality" level, associated say with the inventive step of an invention. The quality level is assumed to be

---

[1] Of course, this is not to say that they are "patentable" – only that patent applications are filed for them. Whether inventions qualify or meet the standards of patentability (as set out in country j's laws) is determined at the patent granting phase.

a random variable, Q, drawn from a negative exponential distribution. Let the cumulative distribution function be as follows.

$$F(q) = \Pr(Q < q) = 1 - \exp(-\psi q), \tag{3}$$

This essentially captures the stylized fact that the distribution of invention quality is skewed: a small percentage of the top inventions account for a large majority of the total value of patent rights.[2] It can be seen from the fact that $F'(q) > 0$ and $F''(q) < 0$. Thus the median quality is less than the mean quality (hence the distribution of Q is positively skewed – or skewed to the right).

The mean inventive step from such a distribution is $1/\psi$. An invention of size q is assumed to augment a firm's productivity by a factor of $\exp(q)$. For example, if A is an index of productivity, the new level of productivity would be $A' = \exp(q) A$. Thus, under this formulation, q is the growth rate of productivity. Firm productivity could be enhanced either because the invention improves production potential or is a cost-saving innovation. We assume that the productivity increase is reflected in firm profits. A tractable formulation is the following.

$$\pi = \pi(q) = \exp(q)\pi_0, \quad \text{where} \quad \pi'(q) > 0,$$

and π denotes the instantaneous flow of profits and π_0 some base flow.

The derivative of π with respect to q under this formulation is of course assumed to be positive. However, it is also necessary to adjust the level of a firm's profits due to imitation activities. Because of imitation or infringement, the returns to inventions are not fully appropriable. In each market or destination j, firms face hazards of imitation. Assume that imitation acts like a tax on profits, and denote by h the rate at which profits can actually be appropriated. Thus net instantaneous profits are

$$\pi = h \exp(q)\pi_0 \tag{4}$$

Where $0 \leq h < 1$. Since $h < 1$, this signifies that the returns are imperfectly appropriated and a value of $h = 0$ denotes that the returns are completely dissipated. However, with patent protection the ability to appropriate increases, depending on the strength of the intellectual property regime. Let θ denote the increase in the rate of appropriation due to patent protection. θ is assumed to be a positive function of the strength of patent rights[3]. Thus with patent protection, the rate at which returns are appropriated equals

---

[2] See Putnam (1996) for some empirical evidence and review of the literature.
[3] If the enforcement of patent rights is completely ineffective or if patent protection is, for some reason, not necessary for the appropriation of investment returns, then θ would be zero.

$h + \theta$, where $0 \leq h + \theta < 1$[4]. The value of a firm equals the presented discounted value of the future stream of profits, and depends on whether or not a firm has a patent. With a patent, the value is

$$V^{PAT} = \int_0^\infty (h+\theta)\pi_0 \exp(q)\exp(-rt)dt = \frac{(h+\theta)\pi_0 \exp(q)}{r} \quad (5)$$

where r is the real interest rate.

Without a patent, the value of a firm is the above expression with θ set to zero:

$$V^{NO\ PAT} = \int_0^\infty h\pi_0 \exp(q)\exp(-rt)dt = \frac{h\pi_0 \exp(q)}{r} \quad (6)$$

Hence the value of patent protection is:

$$\Delta V = V^{PAT} - V^{NO\ PAT} = \frac{\theta \pi_0 \exp(q)}{r} \quad (7)$$

That is, a patent in market j enables the patentee to *purchase* a reduction in the incidence of imitation, the benefit from which is reflected in an increase in firm value. Thus, a firm will seek patent protection if the net benefit of patenting exceeds the cost of filing for protection.

$$\Delta V = V^{PAT} - V^{NO\ PAT} \geq c \quad (8)$$

where c denotes the cost of obtaining a patent (e.g. filing fees, agent fees, and possibly translation fees).

The underlying logic is that inventors have means other than patent protection to appropriate the rewards from their innovation (such as lead times, reputation, secrecy). Thus the value of a patent is the incremental return an inventor can get above and beyond what can be realized by alternative (non-patenting) means.

Equation (8) helps determine which of the source country inventions will be applied for patent protection. A critical threshold quality of inventions can be identified using (7) and (8), namely:

$$q^* = \ln\left(\frac{rc}{\theta\pi_0}\right) \quad (9)$$

---

[4] Note that the increase in appropriability could have been modeled multiplicatively (as $h\theta$). However, the results are qualitatively similar but analytically less tractable.

The more expensive it is to file a patent (i.e. the higher c is), the higher the quality threshold (indicating that only inventions of higher quality are worth patenting). Furthermore, the stronger the patent regime (i.e. the higher θ is), the lower the quality threshold. Thus, not surprisingly, patent rights are more valuable, holding other factors constant, if patent protection is stronger. A higher base flow of profits (π_0) also contributes to a lower critical threshold quality. Firms that in general face larger markets (or produce goods and services that the destination market more highly values) are likely to have a higher base flow of profits.

The number of patents filed can now be determined. Recall the cumulative distribution function F(q). Given a critical threshold quality q*, F(q*) is the fraction of source country inventions that are not patented and 1 - F(q*) = exp(-ψq*) is the fraction that are. Thus, using our notation above, the third term in Eq. (1) is:

$$f_{ij} = \exp(-\psi q^*) = \left(\frac{r c_j}{\theta_j \pi_j}\right)^{-\psi} \quad (10)$$

The subscript j has been brought back to clarify that it is the filing cost and the strength of patent rights of the destination that matter. The base flow of profits π_0 has been renamed π_j to indicate that the profits would be derived from exploiting inventions in the destination market. Note that while these profits are derived from the destination market, they accrue to the firm in source country i. Now, putting it all together by substituting (2) and (10) into (1), yields the prediction for patenting flows from i to j:

$$P_{ij} = \alpha R_i^{\beta_1} S_i^{\beta_2} L_i^{1-\beta_1-\beta_2} \left(\frac{r c_j}{\theta_j \pi_j}\right)^{-\psi} \quad (11)$$

Taking natural logs of both sides of (11) yields:

$$\ln\left(\frac{P_{ij}}{L_i}\right) = \gamma_0 + \gamma_1 \ln\left(\frac{R_i}{L_i}\right) + \gamma_2 \ln\left(\frac{S_i}{L_i}\right) + \gamma_\pi \ln(\pi_j) \quad (12)$$
$$+ \gamma_\theta \ln(\theta_j) - \gamma_c \ln(c_j) + \mu$$

where μ denotes the error term[5].

Empirical measures of these variables are available; for example, data on research and development expenditures and science and engineering

---

[5] Furthermore, the parameters in (11) are functions of previous parameters; i.e. $\gamma_0 = \ln \alpha - \psi \ln r$, $\gamma_1 = \beta_1$, $\gamma_2 = \beta_2$, $\gamma_\pi = \gamma_\theta = \psi$, and $\gamma_c = -\psi$.

personnel can be used for R and S respectively. Patent filing, translation (if any), and search costs can be used for c. An index of patent strength can also be used for θ.[6]

In Eq. (12), the base flow of profits, π_j, depend on the characteristics of the destination market. For example, the size of the destination market (e.g. the markets of the member countries that comprise the EPO) should influence the profitability of commercializing innovations. As a measure of the market size, the real gross domestic product (GDP) of the destination could be employed. For example, the GDPs of the European Patent Convention (EPC) contracting states could be summed (each year) to provide a measure of the market size of the EPC area. Another modification to (12) is to add time subscripts since data on international patenting are available over time. Finally, given the panel data nature, the error term may consist of important individual (country) effects, which may be fixed or random. The individual effect may capture any specific interaction effect that individual source countries have with the EPO.

Thus the equation to be estimated is:

$$\ln\left(\frac{P_{ijt}}{L_{it}}\right) = \gamma_0 + \gamma_1 \ln\left(\frac{R_i}{L_i}\right) + \gamma_2 \ln\left(\frac{S_i}{L_i}\right) + \gamma_\pi \ln\left(GDPC_{jt}\right) \quad (13)$$
$$+ \gamma_\theta \ln\left(IPR_{jt}\right) - \gamma_c \ln\left(c_{jt}\right) + \mu_{it}$$

The error term is motivated by the fact that some profitable inventions fail to be patented while some unprofitable ones are patented. Note that, in (13), j refers to the EPO. Thus, the dependent variable is the natural log of EPO patent applications from country i divided by the labor force in country i.

Equation (13) constitutes a basic model specification for explaining patenting behavior in the EPO. The underlying premise is that inventive and patenting behaviors respond to economic incentives (due to market size, patenting costs, and property rights) and to technological capabilities (such as R&D resources and productivity).

3 Time series regression based approaches

Variants of the model that is developed in Eq. (13) are discussed in Chap. 7, where a time series approach is used to fit the models to available data sets on EPO filings. However, in this book there are several other con-

---

[6] See, for example, Ginarte and Park (1997).

tributions involving time series analyses, and we would now like to introduce methods that are in principle available.

The problem involves setting up a regression analysis based system that is consistent with an economic formulation of the problem. In an empirical regression setting, it is typical that the series to be analysed contains a rather short set of data and so estimation can proceed for only a severely limited number of parameters. It is unlikely that the full economic formulation can be realised in the analysis, since parameters may be essentially collinear and indistinguishable. This is the question of identifiability of a regression model and the basic rule of thumb is to take the simplest alternative that is adequate for the data.

In the self determining approach, the historical development of the patent series itself is used to project the future trend. Usually this is done within the framework of Box-Jenkins based ARIMA methods, state space models or vector autoregression (VAR) models. Chapter 4 introduces these techniques and they appear again in later chapters.

In the predictive approaches, concomitant series of other variables are used to explain the movements in the patent series. Extensions of the above mentioned techniques are generally available to cope with this situation. Chapters 5 to 7 deal with these kinds of models.

Typically, series of pooled national research and development expenditures (R&D) or gross domestic product (GDP) data are used as predictors. Where a model can be developed in which these predictor series act on patents with a lag, forecasts can then be made for the future that are based on predictor values available today. Otherwise, the forecasting problem may just be transferred from one of forecasting the patent series to one of forecasting the values of the concomitant variables. This can in fact make sense if the concomitant variable is something important, such as GDP, where considerable forecasting efforts are available from outside agencies for different purposes.

As with more straightforward linear regression modelling in a conventional statistical setting, simple approaches such as the self determining models can be formulated with ordinary algebra while more complex multivariate models are manageably represented using matrix and vector expressions.

These approaches can be used pragmatically, that is without worrying too much about the underlying economic mechanism. Indeed, for the limited problem of forecasting future numbers of total patent filings, a mechanism may not be needed if the forecasting performance is sufficiently high.

There are of course many developments in time series that have been ignored in the above description, e.g. Nonlinear time series regression methods (Tong 1990). However we suggest that the methods should be kept simple unless evidence emerges that a more complicated formulation is required.

3.1 Econometric methodology – cointegration, vector autoregression, error correction models and count data analysis

Now, we would like to look a little closer at ways that the economic process can be taken into account. The following suggestions involve the VAR methodology but also involve cointegration analysis, error correction models and count data models that otherwise received little attention within the projects that were carried out in the research programme.

The general-to-specific econometric modelling approach advocated by Hendry (1986, 1993, 1995) can be used in the modelling and forecasting of patent application filings. However, to be consistent with our stated aim in the last section to keep modelling simple, we suggest that even the initial general specification should not be too elaborate. The general-to-specific modelling approach is a relatively recent strategy used in econometrics. It attempts to characterize the properties of the sample data in simple parametric relationships that remain reasonably constant over time, account for the findings of previous models, and are interpretable in an economic and financial sense. Rather than using econometrics to illustrate theory, the goal is to "discover" which alternative theoretical views are tenable and test them scientifically. The general-to-specific approach starts with over-parameterized models (in terms of lags and variables) and through specification tests reduces to a parsimonious representation.

Consider for example a plan to develop models to describe patenting at the EPO by applicants from various countries using this approach. The initial models can rely on a database that would at a minimum include the filings from various countries at the EPO, domestic filings, R&D measures, and real GDP. There may be a long-run relationship in the level or accumulated R&D effort and patenting activity. Basic scientific developments and technological progress can lead to important or break-though innovations, which produce large flows of patent activity and filings. Also, the size of the domestic market for innovation can influence the propensity to file a patent at the EPO. The mechanism of the patent system, which involves a priority forming first filing at one office that is followed by subsequent filings at other offices, suggests that there is a lag between the domestic filings and subsequent filings at the EPO. Real GDP captures the demand for new products, processes, and services. In the short run there can be another relationship between year-to-year changes in R&D effort, which provides incremental technological improvements, and modifications that affect patent application filings in the short run. In the next or final stage the national level models can be combined in either a panel data framework or in a system of equations.

Once the variables have been selected, the process begins with examining the time series properties of the data. The (statistical) implication of a series with a unit root is that it is non-stationary or integrated of order one, I(1). There is no tendency towards a mean value and the standard error is not defined. We can look at the change of these series or their growth rates and these will be stationary with a defined mean and standard error. Simple univariate forecasting models can be developed based on the time series properties. These models can be used as a comparison for the econometric models. In some cases, macroeconomic series with unit roots tend to "trend" together or are cointegrated. They might not synchronously rise and fall every year, but in the long run they move together. Examples of this include exchange rates and interest rate differentials, consumption expenditures and (disposable) income, levels of sales and inventories or capital stock, short run and long-run interest rates, and patenting activity or filings and R&D effort. These examples represent equilibrium relationships that are consistent with economic theory and intuition. The standard test procedure follows the multivariate cointegration approach of Johansen (1988).

The procedures begin with a multivariate specification. Patent filings can be represented as the flow of new knowledge which is a function of the stock of knowledge derived from previous patents. The labor effort devoted to knowledge production can be represented by R&D effort and real GDP. A preliminary national level VAR model might look like:

$$\begin{bmatrix} EPOFil_t \\ RD_t \\ RGDP_t \\ DOMFil_t \end{bmatrix} = A(0) + A(L) \begin{bmatrix} EPOFil_{t-1} \\ RD_{t-1} \\ RGDP_{t-1} \\ DOMFil_{t-1} \end{bmatrix} + B(L)X_t + e_t$$

where EPOFil are the filings at the EPO, RD is a measure of R&D effort, RGDP is real gross domestic product, and DOMFil are first filings nationally or the size of the domestic innovation market. Patent application filings are an indicator for knowledge production. Country specific subscripts are omitted for simplicity here, but specific models may be developed for each country. A(0) is a set of deterministic variables like constant, trend, or dummies for changes in patenting rules. The specification and interpretation of these variables will be discussed shortly. A(L) is a lag polynomial operator (see also Chap. 4 and Chap. 6), X are other exogenous variables, contemporaneous and lagged depending on the conformable operator B(L). The term e represents a vector of disturbances with means of zero, constant variance covariance matrix, and serially uncorrelated disturbances.

The long-run properties of trending variables can be used to explain growth or speeds of adjustment to equilibrium over time through an error correction (ECM) model. A special case of the ECM is the partial adjustment model. The ECM captures both the short-run dynamics and the long-term trends in patent filings. The specification is in growth rates or first differences in natural logarithms. This approach avoids the resulting nonsense or spurious regression problem when the data series are I(1). Research by Diebold and Lutz (2000), and Clements and Hendry (2000) suggest that the cointegration testing and possibly an error correction model is the appropriate methodology to use for estimation and forecasting purposes.

Exogeneity is an important conceptual and empirical issue for inference analysis, forecasting, and policy analysis in the presence of cointegration. Ericsson and Irons (1994) discuss these issues and the appropriate empirical tests. Joutz and Maxwell (2002) have applied these techniques in modelling and forecasting high yield bonds. The factorization or conditioning of models following these tests suggests the appropriate specification of the forecasting models.

Cointegration tests are a multivariate form of integration analysis. Individual series may be I(1), but a linear combination of the series may be I(0). The error correction model is a generalization from the traditional partial adjustment model and permits the estimation of short-run and long-run elasticities.

An approach to this is based on the findings of Nelson and Plosser (1982), in which many macroeconomic and aggregate level series are shown to be well modeled as stochastic trends, i.e. integrated of order one, or I(1). Simple first differencing of the data will remove the nonstationarity problem, but with a loss of generality regarding the long run "equilibrium" relationships among the variables.

Engle and Granger (1987) solve this filtering problem with the cointegration technique. They suggest that if all, or a subset of, the variables are I(1), there may exist a linear combination of the variables which is stationary, I(0). The linear combination is then taken to express a long-run "equilibrium" relationship. Series which are cointegrated can always be represented in an error correction model. The error correction model is specified in first differences, which are stationary, and represent the short run movements in the variables. When an error correction term is included in the model, the long-run, or equilibrium, relations in patent activity are accounted for. Lags of the independent and dependent variables may be included to capture additional short- and medium-term dynamics of patenting activity. The advantage of the first difference model is that the specification is stationary so that estimation and statistical inference can be performed using standard statistical methods. The contemporaneous coefficients are interpreted as short run elasticities.

The vector(s) obtained in the cointegration analysis represent the long-run relationship among the variables. To model patent filing or activity more generally, however, a short-run error correction model is employed. The error correction framework models the variables in differences, and then the coefficients on the differenced variables correspond to short-run elasticities. The model furthermore contains an error correction term, ECM. This term is obtained from the long-run relationship and expresses deviations in patent filing (knowledge production) from its long-run mean. The coefficient in front of the ECM term measures the speed of adjustment in current consumption to the previous equilibrium demand value. The model in its most general form is as follows:

$$\Delta y_t = \alpha + \Sigma \delta_i \Delta y_{t-i} + \Sigma \beta_i \Delta x_{t-i} + \gamma ECM_{t-1} + e_t \quad i = 1,\ldots,t$$

where y are the dependent variables, x is a vector of independent variables and ECM is the error correction term.

Once these first round models are built, the effort can turn to simultaneous modelling of the patent filings at the EPO. There are several approaches that can be considered. These will depend on data quality, sample size, and results from the preliminary models.

One approach is to use dynamic panel data techniques. In a panel data setting, there are time-series observations on multiple countries. We can denote the cross-section sample size by N, and, in an ideal setting, have t=1,...,T time-series observations covering the same calendar period, a balanced panel. In practice, it often happens that some cross-sections start earlier, or finish later. When T is large, N small, and the panel balanced, it is then possible to use the simultaneous-equations modelling procedures. If T is small, and the model is dynamic (i.e. includes a lagged dependent variable), the estimation bias can be substantial. See Nickell (1981). Methods to cope with such dynamic panel data models include the GMM-type estimators of Arellano and Bond (1991), Arellano and Bover (1995), Blundell and Bond (1998). Li et al. (1997) have used shrinkage estimation techniques for ECM modelling of energy demands.

The second approach uses the simultaneous-equation modelling framework. The approach may start with the unrestricted VAR and impose conditions on the "exogeneity" of the variables and the variance-covariance matrix.

It should be noted that patent filings are count data in nature (e.g. non-negative integers). In that regard, a complementary approach is to use count data analysis.[7] Under this approach, the researcher would work directly with the raw data. Patent counts would not be transformed (for example, through differencing, expressing filings as ratios of other variables like labor or GDP, and/or taking logarithms); otherwise by definition they would no longer be count data. In working with raw data, however, we would encounter issues concerning the stationarity of the variables; the patent counts could be integrated of order 1. Secondly, the number of filings is generally non-zero and generally "large" in value for most countries during most periods. Relatively few zeros or small numbers (e.g. under 20) are encountered. The advantages of using count models are therefore not especially great for international patent filing data. Nonetheless, future work could better incorporate count data modeling in patent forecasting analyses.

4 Conclusions

Some of the approaches described in Sect. 3.1 have been applied to modeling the development of knowledge using USPTO patent data (Abdih and Joutz 2005). This seems to be a good path to follow in future approaches to EPO data. It would be possible to take even deeper account of econometric theory in formulating time series based regression models. For example, a commonly used approach in macro-economic model building is to posit a model including a whole set of contributory variables, to test and isolate for exogenous versus endogenous effectors, and then to isolate combinations of endogenous effectors in reduced form equations that can be used for the actual fitting (Favero 2001). This approach is of importance when looking at frequently measured sets of correlated variables (e.g. interest rates, GDP, etc.), but may be more suitable for considering policy implications and intervention analysis, rather than for forecasting numbers of patent filings from historical annual counts.

The knowledge processes that are considered are amenable to varying interpretations in terms of modeling approaches. It remains an open question as to whether a detailed examination of knowledge production is appropriate when constructing a forecasting model for budgetary purposes. This is because the data that are likely to be available for modeling may lack the richness required for uncovering the processes that are described.

---

[7] See the seminal work of Hausman et. al. (1984).

Nevertheless, any simplified model that is used for regression analysis should at least be consistent with an accepted underlying mechanism of knowledge production.

However, the studies in the following chapters do not generally take up on the suggestions made here. Differencing as a form of dealing with cointegration is approached via Box-Jenkins ARIMA methodology in Chap. 4 and Chap. 6, while in Chap. 5 Blind uses an alternative approach that uses a trend effect as an additional variable, and in Chap. 7 an extension of the approach from Sect. 2 of this chapter is used. The various approaches in the following chapters are practical and useful methods for the problem at hand.

Chapter 4

An assessment of the comparative accuracy of time series forecasts of patent filings: the benefits of disaggregation in space or time

Nigel Meade

Tanaka Business School, Imperial College London, London, UK

Module: E
Software: STAMP, TSP

1 Introduction

This work is concerned with methods for forecasting the filing of patents and was carried out in conjunction with the European Patent Office. The filings data were subdivided by:

- Blocs – European Patent Convention countries, Japan, US and the Rest of the World.
- Industries – main Fields of Technology according to headings A–H of the International Patent Classification (WIPO 2000).

The issues addressed are: the benefits of multivariate models versus univariate ones in exploiting any correlations between the filings in different blocs or industries; the effect of aggregation over time (from monthly to annual data) and the effect of aggregation by bloc or by industry on the accuracy of the forecast of total EPO filings. Two approaches are used: the ARIMA framework and the dynamic linear model (DLM) in both univariate and multivariate modes.

The main results are: monthly data does tend to provide greater accuracy in annual forecasts; there are no significant benefits to be gained by multivariate modelling and no significant benefits are found from aggregating over blocs or industries. There are benefits from using monthly data, rather than annual data. The best modelling approaches are, for monthly data, the univariate dynamic linear model; for annual data either the univariate ARIMA or DLM could be used. The recommended forecasting approach provides a benchmark against which other forecasts drawing on different data sources can be compared.

The filing of an application for a patent is the first step in achieving protection for intellectual property. The three major patent offices are the European (EPO), the Japanese (JPO) and the US (USPTO) offices. The examination of each application is a labour intensive process, involving technological and legal expertise. The motivation for this study of the forecasting of patent filings with the EPO is that the forecasts are a prerequisite for manpower planning at an aggregate scale and at the level of availability of expertise in different technologies. The data used in this study was made available in 2002 as part of an EPO research programme. The programme looks at five different approaches to forecasting patent filings, see Chap. 2. These approaches include: survey methods; micro-level studies of firms' patenting practice; effects of inter-firm competition on patenting behaviour; consideration of the flows of filings between patent offices; time series modelling of aggregate filing data. It is this last topic that is dealt with here.

The levels of aggregation at which patent filings are analysed are:

- Bloc level – filings are from the EPC contracting states, US, Japan and the rest of the world.
- Industry level – nine industry groupings are used.
- Total filings within the EPO.

The objective is to identify the most accurate means of forecasting filings within these categories. This analysis of patent filings, at several levels of aggregation, uses two time series modelling frameworks.

The basic methodologies used are:

- ARIMA model.
- Dynamic linear model.

Three central issues are:

- Is there information in the filings within blocs or industries that would lead to increased forecasting accuracy via the use of multivariate forecasting models?
- Does disaggregation over blocs or industries lead to improved forecasting?
- Does disaggregation over time lead to improved forecasting? Does the use of monthly data, rather than annual data, lead to greater forecasting accuracy?

In other words, is it better to forecast the filings from the blocs (or industries) individually and then consolidate them; or to simply forecast the consolidated figures?

Out of sample forecasting accuracy over a one year ahead horizon is the criterion upon which model performance is judged.

This chapter is divided into the following sections: a description and exploratory analysis of the data; a discussion about the forecasting methods to be used; the application of the forecasting methods to the data; an analysis of forecasting accuracy; forecasting accuracy over a longer horizon followed by a summary and conclusion.

2 Data description

The data provided represent filings via two mechanisms: direct applications to the EPO (euro-direct) and via the Patents Cooperation Treaty (euro-PCT-IP). For further details, see WIPO (2006a).

Data are available on both a monthly and a yearly basis. For the analysis, the data were consolidated into bloc or industry series shown in Table 4.1. The relative proportions of filings in the different world blocs and industry groups at the EPO are shown in Fig. 4.1. Industry is used as shorthand for the main field of technology according to the International Patent Classification, see WIPO (2000).

Table 4.1. The countries / blocs and industries for which data are available

Bloc	Code
European Patent Convention Signatories	EPC
Japan	JP
United States	US
Rest of world	ROW
Total	EPO
Industry	Code
Human Necessities	A
Performing Operations/Transport	B
Chemistry; Metallurgy	C
Textiles; Paper	D
Fixed Constructions	E
Mechanical Engineering	F
Physics	G
Electricity	H
Other	other

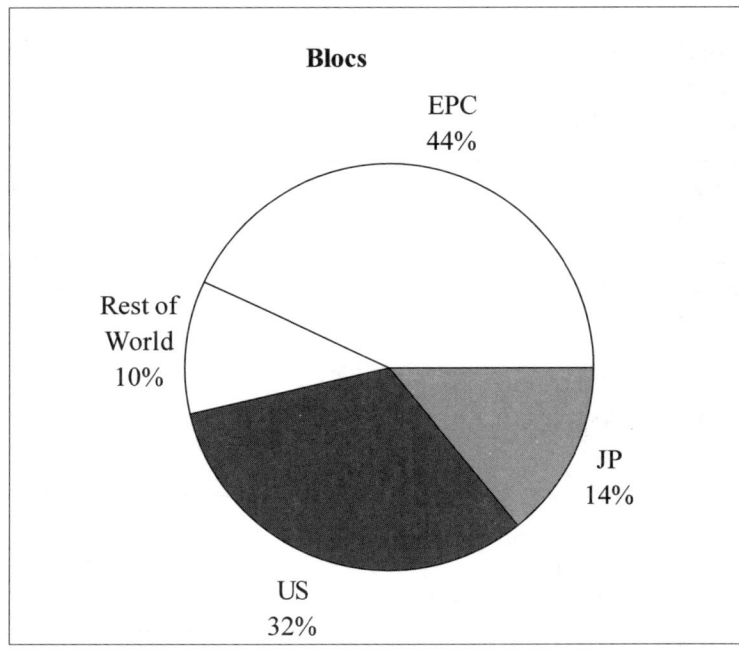

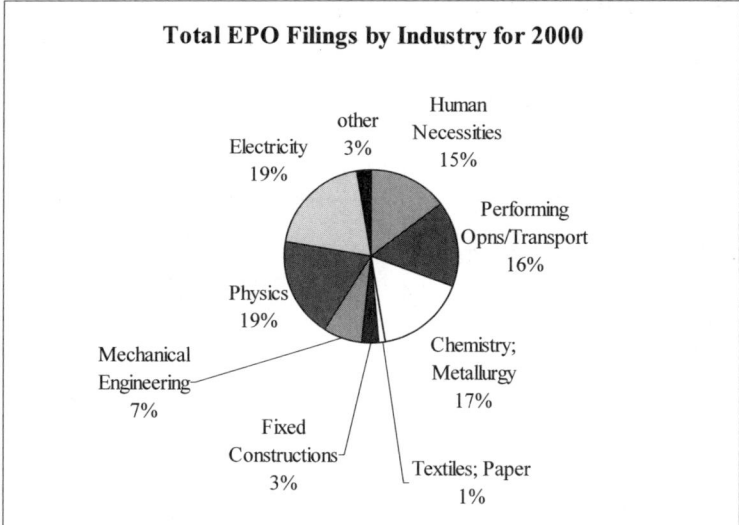

Fig. 4.1. Proportions of filings by bloc and by industry (IPC main fields)

An important issue in this project is the extent to which filings from one bloc or industry affect filings in another bloc or industry as this will impact on the performance of multivariate models. The filings data over the period available has trended upward throughout with the occasional hesitation. The data are graphed in Fig. 4.2, giving annual filings by blocs and total EPO filings; and in Fig. 4.3, giving the filings by industry groupings.

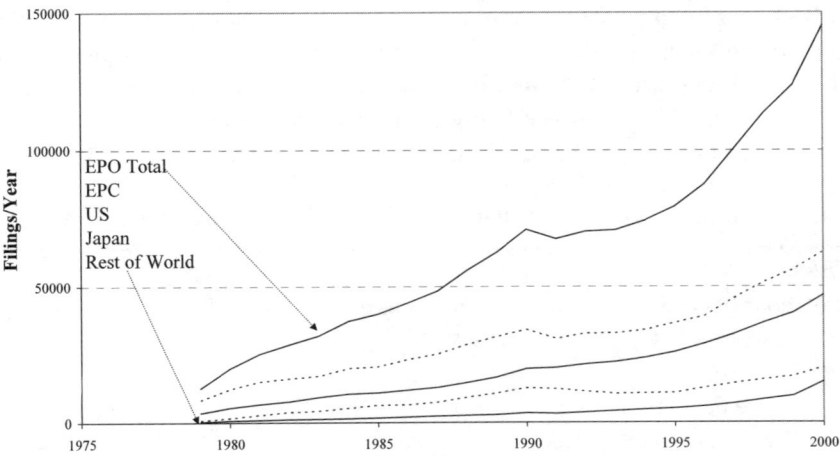

Fig. 4.2. Annual filings for EPO blocs

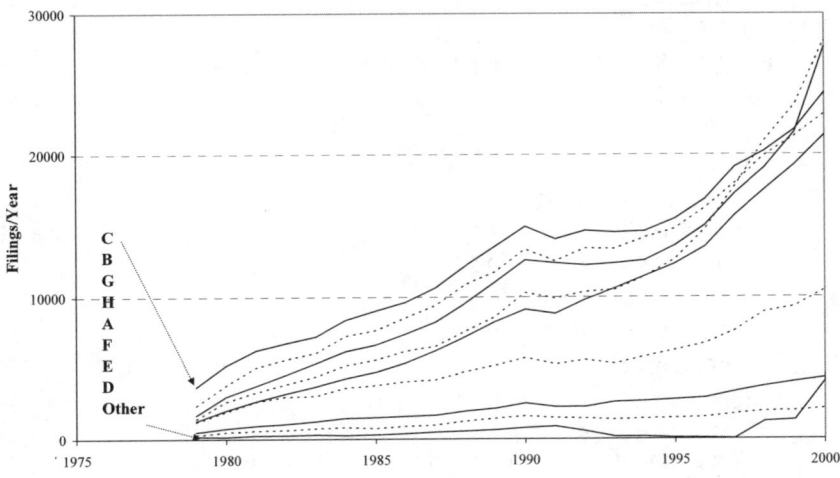

Fig. 4.3. Annual filings by industries

Here we are interested in any evidence of one industry or bloc *leading* another. This is different to the patent family approach, see Hingley and Park (2003), where the progress of filings at different offices is modelled. The strong trend apparent in the data (X_t) means that filings by bloc or industry will be strongly correlated simply because of the common trend. For correlations to be meaningful, the data need to be stationary, that is their mean and variance should not change over time. Thus, it is more informative to look at the correlation between annual changes in filings $(X_t - X_{t-1})$ as these changes are stationary. The correlations are tabulated in Table 4.2.

Table 4.2. Correlations between annual changes in filings using annual data

By Bloc				
Correlations between coincident annual changes				
	EP	JP	US	ROW
EP	1.00			
JP	0.66	1.00		
US	0.92	0.62	1.00	
ROW	-0.67	-0.22	-0.78	1.00

Inter-bloc correlations are significant, correlations with rest of world and other blocs either not significant or negative
No significant correlations were found at other lags

By Industry									
Correlations between coincident annual changes									
	A	B	C	D	E	F	G	H	other
A	1.00								
B	0.95	1.00							
C	0.95	0.96	1.00						
D	0.86	0.92	0.91	1.00					
E	0.92	0.93	0.91	0.87	1.00				
F	0.88	0.94	0.91	0.89	0.85	1.00			
G	0.79	0.76	0.85	0.72	0.73	0.80	1.00		
H	0.90	0.88	0.90	0.79	0.83	0.91	0.93	1.00	
other	-0.89	-0.84	-0.82	-0.77	-0.81	-0.70	-0.51	-0.64	1.00

Virtually all correlations significant. No significant correlations were found at other lags

Unfortunately, there was no evidence of a lagged relationship between blocs or industries; that is, there was no evidence that a change in one section preceded change in another. The different filings groups seem to respond to the same stimuli at the same time.

In order to examine the possible presence of a lagged relationship further, the exercise is repeated using monthly data, this extra 'definition' in the data might reveal something hidden in the annual data. In order to achieve stationarity, the seasonal effect is removed by 'seasonal differencing'. The data used are $(X_t - X_{t-1} - X_{t-12} + X_{t-13})$ and the correlations are shown in Table 4.3.

Table 4.3. Correlations between monthly changes in filings using monthly data

By Bloc

Correlations between coincident monthly changes

	EP	JP	US	ROW
EP	1.00			
JP	0.73	1.00		
US	0.66	0.64	1.00	
ROW	0.02	0.21	0.00	1.00

Correlations with rest of world and other blocs either not significant or negative. No significant correlations were found at other lags

By Industry

Correlations between coincident monthly changes

	A	B	C	D	E	F	G	H	other
A	1.00								
B	0.79	1.00							
C	0.83	0.84	1.00						
D	0.62	0.68	0.64	1.00					
E	0.61	0.74	0.63	0.49	1.00				
F	0.81	0.83	0.82	0.64	0.68	1.00			
G	0.82	0.88	0.87	0.65	0.68	0.81	1.00		
H	0.79	0.80	0.86	0.67	0.60	0.79	0.90	1.00	
Other	-0.21	-0.08	-0.09	-0.02	-0.13	-0.15	-0.12	-0.10	1.00

Virtually all correlations significant. Several slightly significant correlations were found at other lags

The correlations for monthly changes are typically lower than for the annual data. This is due to the greater stringency that is asked for here, because changes have to happen not just in the same year but in the same month. However, there was still little evidence of lagged effects.

3 Review and description of forecasting methods

The forecasting methods discussed here are extrapolative methods, this means that the information set used for forecasting is the history of the relevant variable or variables. Extrapolation in its widest context is discussed by Armstrong (2001). The three central issues mentioned in the introduction will be discussed one by one here. Firstly, the value of using a multivariate model rather than a set of univariate models will be examined. If there is correlation between the set of series being forecast, then the multivariate model may be expected to capture this extra structure and hence produce more accurate forecasts. Preez and Witt (2003) compared univariate and multivariate ARIMA and state space model based approaches to forecasting international tourism (the dynamic linear model is a particular implementation of a state space model). The data were the numbers of visitors to the Seychelles from four European countries. They found that their multivariate (state space) model was uniformly least accurate and that univariate ARIMA was, on average, most accurate. Their findings demonstrate that the use of multivariate models can involve loss of accuracy, perhaps due to constraints in their structure, as well as the opportunity of increased accuracy.

Secondly, the issue of cross-sectional aggregation will be examined. Aggregation across the blocs of the EPO is a form of spatial aggregation, a topic addressed by Miller (1998). He points out that the econometric literature is mainly concerned with the effects of aggregation on the density functions of parameter estimates rather than forecasting accuracy. In his study of forecasting economic variables, such as unemployment, in regions within a US state, he found no real difference in accuracy between forecasts using disaggregated and aggregate data. Individual industry forecasts are analogous to *bottom up* forecasts in the context of business forecasting, while the alternative is the *top down* forecast where an aggregate forecast is sub-divided. In this context, Schwartzkopf et al (1998) and Dangerfield and Morris (1992) found that disaggregated (bottom up) forecasts tended to be more accurate than aggregated forecasts.

The third issue raised was the effect of temporal aggregation, from the patent filings perspective using annual data when monthly data is available. The issue of time deformation is discussed by Stock (1988). He considers the situation where there is an "operational time scale" rather than a calendar scale. Essentially, events influence the rate of change of the process being studied. Although a time transformation is not considered feasible here, it is possible that the evolution of the patent filing process is not fully summarised by annual data and that monthly data provides more useful information. This argument influenced Funke (1990), who used monthly Vector AutoRegression (VAR) models to "capture the current economic outlook as quickly as possible" when forecasting industrial production in OECD countries. Rossana and Seater (1995) find evidence of a substantial loss of information when temporal aggregation of monthly or quarterly data to annual data is performed. The information lost described low frequency cyclical variation. Perhaps as a consequence of this, they found that the aggregated annual data showed more long run persistence than the underlying higher frequency data.

In order to try and illuminate these issues, two different time series frameworks were used for modelling the data. These are ARIMA modelling and the dynamic linear model. Software for the analysis is Time Series Processor (TSP[1]) and Structural Time series Analyser, Modeller and Predictor (STAMP[2]) respectively. These frameworks will both be used in univariate and multivariate modes to address the first issue. To address the third issue, models will be derived using both monthly and annual data. The second issue of cross-sectional aggregation will be examined by aggregating over blocs and industries.

ARIMA modelling is well described in Box and Jenkins (1976) or Pankratz (1983). For the stationary process, Z_t, the general ARMA process has p autoregressive terms and q moving average terms.

$$Z_t - \phi_1 Z_{t-1} \ldots - \phi_p Z_{t-p} = a_t - \theta_1 a_{t-1} \ldots - \theta_q a_{t-q} \qquad (1)$$

The variable, a_t, is a white noise term where:

$$a_t \sim N(0, \sigma^2) \quad \text{and} \quad \rho(a_t, a_{t-k}) = 0 \quad \text{for} \quad k \neq 0$$

We use the conventional notation where B is the backward difference operator, B, where

$$BX_t = X_{t-1} \quad \text{so} \quad B^d X_t = X_{t-d} \quad \text{and} \quad (1-B)X_t = X_t - X_{t-1}$$

In this context, X_t represents the number of patents filed at time t, and Z_t represents a stationary transformation of X_t, after finding a suitable value

---

[1] See http://www.tspintl.com.
[2] See http://stamp-software.com.

for d. The estimation of the coefficients is achieved by maximum likelihood, the necessary first step is the identification of p and q. This identification is achieved by using an information criterion. The criterion balances the marginal value of increasing the number of parameters; say by increasing p by one, against the reduction in the variance of the noise term, a_t. The underlying intuition is that the criterion allows the extraction of the underlying structure of the data without over-fitting the model. In the following analysis, the values of p and q in ARMA(p,q) were chosen by minimising Schwarz's Bayesian Information Criterion (see, for example, Mills, 1999). The criterion is:

$$BIC = \ln(\hat{\sigma}^2) + m\frac{\ln(T)}{T} \qquad (2)$$

where there are T observations used to calculate the standard error $\hat{\sigma}$, an estimate of σ where $V(a_t) = \sigma^2$ and $m = p + q$ coefficients estimated.

Monthly data are likely to exhibit seasonality, a pattern that tends to repeat itself every year. The most common ARIMA model for handling monthly data is called the multiplicative model. It is a combination of two models, one representing within year behaviour, with L observations per year, and one representing annual behaviour. This can be denoted by ARIMA(p,d,q)x(P,D,Q), and its equation is:

$$(1-B^d)(1-B^{DL})(1-\phi_1 B ... -\phi_p B^p)(1-\Phi_1 B^L ... -\Phi_P B^{PL})X_t = \qquad (3)$$
$$(1-\theta_1 B ... -\theta_q B^q)(1-\Theta_1 B^L ... -\Theta_Q B^{QL})a_t$$

The identification and estimation procedure is analogous to that for non-seasonal series.

The multivariate version of ARIMA modelling used here is VAR. The model is:

$$Z_t = \mu + \Phi_1 Z_{t-1} + ... \Phi_p Z_{t-p} + \varepsilon_t \qquad (4)$$

where Z_t is now a vector of length k of observations of k stationary time series and Φ_i (i = 1,... p) are $k \times k$ matrices of coefficients, μ is a vector of constants and ε_t is a vector of white noise terms with zero mean and a non-singular covariance matrix. The vector time series, Z_t, is formed by taking first differences of the raw data. The order of the process, p, is normally determined by likelihood ratio tests or by information criteria. However, in the case of yearly data, shortage of observations dictates the maximum value of p.

The dynamic linear model is founded on the simple hypothesis that a time series can be decomposed into a level, a trend (called a slope in the software used), seasonality (if relevant) and a random component. The model is a 'state space' model in that it is described by an observation equation and several more equations describing the evolution of the states.

The observation equation is:

$$X_t = \mu_t + \varepsilon_t \qquad \varepsilon_t \sim N\left(0, \sigma_\varepsilon^2\right) \qquad (5)$$

where the level, μ_t, is defined by

$$\mu_t = \mu_{t-1} + \beta_{t-1} + \eta_t \qquad \eta_t \sim N\left(0, \sigma_\eta^2\right) \qquad (6)$$

and the trend, β_t, is defined by

$$\beta_t = \beta_{t-1} + \zeta_t \qquad \zeta_t \sim N\left(0, \sigma_\zeta^2\right) \qquad (7)$$

The random changes to level, η_t, and trend, ζ_t, and the random disturbances, ε_t, about the level are each independent of each other. As a new observation becomes available, estimates of the current level and trend are revised. The forecast for k periods ahead is the current estimate of the level plus k times the current estimate of the trend, i.e.

$$E(X_{t+k} | X_t) = \hat{\mu}_t + k\hat{\beta}_t \qquad (8)$$

This formulation is also called an unobserved components model. If the noise terms identified earlier, η_t for level and ζ_t for trend, are each related linearly to the random disturbance, ε_t, then the dynamic linear model is said to be a single source of error model. In this case, the model described is identical to the ARIMA(0, 2, 2) model. This type of parallel is discussed by Gardner and McKenzie (1988). A dynamic linear model with multiple sources of error as defined above cannot be well represented by the ARIMA structure simply because it only admits one source of error. This is discussed in Meade (2000). A framework of dynamic linear models with a single source of error is given by Hyndman et al (2002). This covers additive trend (the case used here), multiplicative trend and an absence of trend. This framework also includes seasonal components that are modelled as either additive or multiplicative terms.

STAMP, the dynamic linear modelling software, offers several options for modelling seasonal variation. Seasonal factors can be fixed constants (estimated from the data) or they can be modelled as stochastic trigonometric factors. The former alternative is equivalent to the additive seasonal factor in Hyndman et al's framework of state space models. However, the latter alternative was used for this analysis, because a model of the whole seasonal profile is used rather than a set of disjoint factors. The seasonal variation within a year is represented as a weighted sum of sine and cosine waves (the fitting procedure is a form of Fourier analysis). The additive seasonal factor for month m (=1, 12) is represented below.

$$\text{Seasonal factor}_m = \sum_k \beta_{S,k} \sin\left(\frac{2\pi mk}{12}\right) + \sum_k \beta_{C,k} \cos\left(\frac{2\pi mk}{12}\right) \tag{9}$$

The Fourier series contains as many terms as necessary to capture the seasonal profile, subject to there being no more terms than months. In other words, this representation is usually more parsimonious (requires fewer estimates) than individual monthly factors.

The multivariate version of the dynamic linear model is a form of the seemingly unrelated time series equations (SUTSE) model. SUTSE implies a set of linear models linked only by their disturbances, which may be correlated. If the linear disturbances are uncorrelated, then the procedure is equivalent to separate linear models. The greater the correlation, the greater the efficiency gain in estimation using this approach. Note that multivariate dynamic linear model implies SUTSE, not vice versa.

SUTSE is a generalisation of (5,6 and 7) where vectors replace the dependent variable and the level and trend parameters. The variances of univariate disturbances, $\sigma_\varepsilon^2, \sigma_\eta^2, \sigma_\varsigma^2$ become $k \times k$ covariance matrices, $\Sigma_\varepsilon, \Sigma_\eta, \Sigma_\varsigma$. A homogeneous model is used, where the covariance matrices of the disturbances are assumed proportional to one another. That is $\Sigma_\eta = q_\eta \Sigma_\varepsilon$ and $\Sigma_\varsigma = q_\varsigma \Sigma_\varepsilon$ where q_η and q_ς are scalar parameters.

4 Application of forecasting methods

For the analysis, all the data series are split into two regions. In the estimation region, the data are used for model identification and parameter estimation only. The remaining data are used for out of sample forecasting. Since forecasting is done recursively on an annual basis, after a year has been forecast, that year's data is then included for parameter estimation for the forecast of the following year. The estimation and forecast regions are defined in Table 4.4.

Table 4.4. Definition of estimation and forecast regions

		Yearly	Monthly
Total availability	Start date	1979	1978 June
	Finish date	2001	2001 Dec
Number of observations		23	283
Data used only for estimation	Start date	1979	1978 June
	Finish date	1994	1994 Dec
Number of observations		16	199
Data used for out of sample forecasting and subsequent estimation	Start date	1995	1995 June
	Finish date	2000	2000 Dec
Number of observations		6	72

Although data were available for 2001, their quality was suspect and these data were not used in the evaluations of forecast accuracy.

Once the model has been identified for each series, the forecasting process is carried out recursively. Forecasts from the model estimated using data in the estimation region are prepared for up to a year ahead. This additional data is then used for parameter re-estimation; forecasts are prepared for a year ahead and so on. In summary, forecasts with a maximum horizon of one year are prepared sequentially for six origins at annual intervals.

These sets of forecasts provide accuracy information on a multi-origin and multi-horizon basis, as recommended by Fildes (1992). The measures of accuracy used are root mean square error (rmse) and mean absolute percentage error (mape). Both of these measures are calculated over different origins $i = 1, \cdots, I$, over horizons $h = 1, \cdots, H$, (for monthly data only in this section, multiple annual horizons are examined in Sect. 4) and overall. If the forecast origin is $T + (i - 1)M$ (where: for annual data $T = 16$, $M = 1$, $I = 5$ and for monthly data $T = 199$, $M = 12$, $I = 5$); then the h step ahead forecast $\hat{Y}_{T+(i-1)M,h}$ is an estimate of the observation $Y_{T+(i-1)M+h}$ and the error is $e_{i,h} = Y_{T+(i-1)M+h} - \hat{Y}_{T+(i-1)M,h}$. The percentage error is

$$p_{i,h} = \frac{100 e_{i,h}}{Y_{T+(i-1)M+h}}.$$

The measures of accuracy are given in Table 4.5.

Table 4.5. Definitions of accuracy measures used

	root mean square error (rmse)
For origin i	$RMSE_{Origin}(i) = \sqrt{\dfrac{\sum_{h=1}^{M} e_{i,h}^2}{M}}$
For horizon h	$RMSE_H(h) = \sqrt{\dfrac{\sum_{i=1}^{I} e_{i,H}^2}{I}}$
Overall O	$RMSE_O = \sqrt{\dfrac{\sum_{h=1}^{M}\sum_{i=1}^{I} e_{i,h}^2}{I\,M}}$
	mean absolute percentage error (mape)
Overall O	$MAPE_O = \dfrac{\sum_{h=1}^{M}\sum_{i=1}^{I} |p_{i,h}|}{I\,M}$

As a general rule, the rmse is reported as the prime measure of accuracy and mape is used as a supporting figure, often for comparison between series.

Accuracy over a horizon of one year is used as the main criterion for comparison between models for two reasons. Firstly, a year is the shortest common horizon for the annual and monthly data sets. Secondly, the coefficients of both modelling procedures are estimated to best describe the relationship between an observation at time *t* and those observations available at time *t-1*. Each model has the objective of minimising the one period ahead error variance.

The focus in this study is on a one-year forecasting horizon, because in practice the forecasting exercise will be repeated at intervals of no more than a year. Some analysis of forecasts for horizons of up to four years is given in Sect. 6.

In the analysis, root mean square errors are used to estimate the one-step ahead error standard deviation.

The application of the forecasting methods is described in the following sub-sections.

4.1 ARIMA univariate (annual data):

The values of p and q, for ARIMA(p, d, q), were chosen using the Bayesian Information Criterion. In two (monthly) cases, non-differenced models were used. Firstly, the results of forecasting annual filings are discussed, then those for monthly data.

Aggregate forecasts of total EPO filings are produced; these forecasts are the sum of forecasts of the components of the total EPO filings. These components are either the four blocs (EPC, Japan, US, Other countries) or industries (IPC main Section headings A – H). A comparative analysis of this type produces a large volume of results, the policy adopted here is to first give example tables for the ARIMA models using annual data; subsequently summary results are given. The example tables give an idea of the relative difficulty of forecasting the different series, the summary tables show the relative overall accuracy between methods.

For the four blocs (see Table 4.6), the US has the lowest mape; the rest of the world has the highest. Forecasts labelled origin 1 are for 1995, based on data up to 1994. Forecasts labelled origin 2 are for 1996, based on data up to 1995, etc. The forecast based on aggregated blocs gives a similar rmse to the direct forecast of total EPO filings. The rmses associated with Origin 6 are typically high. These relate to the forecasts made at the end of 1999 for 2000. Total filings increased by 15% from 1999 to 2000 in contrast with the 7% growth experienced over the last ten years, 1990 to 2000. All four univariate forecasts of total EPO filings are shown in Fig. 4.4.

Table 4.6. Summary accuracy of ARIMA forecasts of total EPO annual filings forecasts by blocs

		EPC	JP	US	Rest of World	Blocs aggregated	Total EPO
	Std Error	1900	849	722	224		3211
Accuracy	1	895	97	659	52	1405	1129
by	2	575	1780	1759	310	4424	3754
origin	3	4706	355	2221	841	8124	9750
(rmse)	4	3927	103	983	749	5762	2352
	5	539	144	278	172	845	1839
	6	1863	2260	2796	3833	10753	10593
Overall rmse		2660	1186	1699	1638	6286	6212
Overall mape		4.29	4.91	4.11	8.91	4.68	4.38

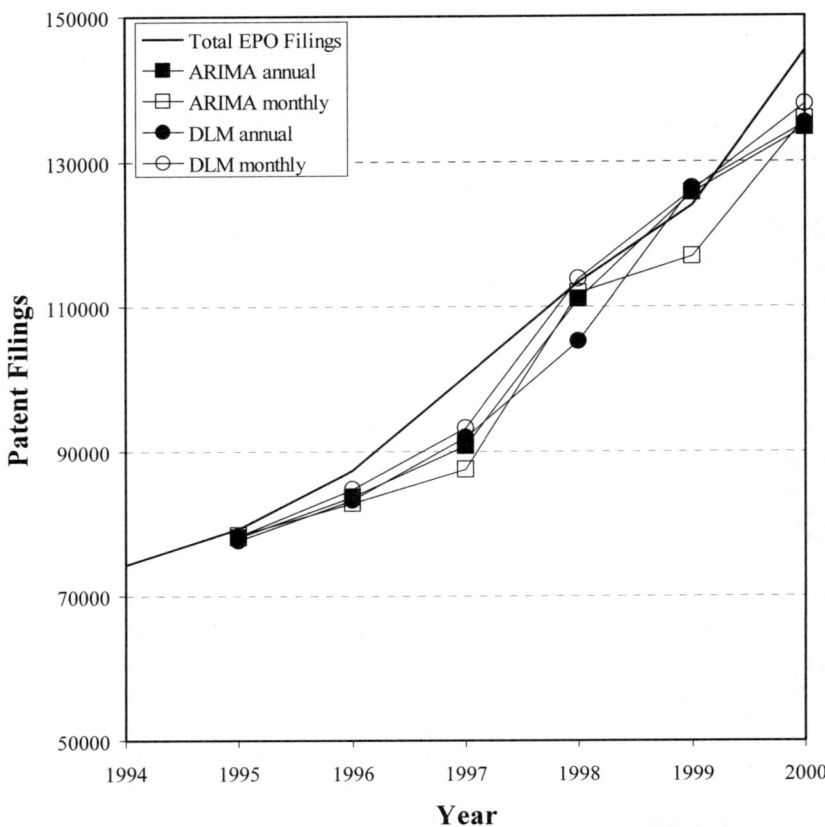

Fig. 4.4. One year ahead forecasts of total EPO filings for origins 1994 to 1999 from the four univariate models

The results of examining filings on an industry basis for the EPO are summarised in Table 4.7. industries G (physics) and H (electricity) are the most difficult to forecast with both high root mean square errors and high mean absolute percentage errors. The residual category 'other' has a high mape and high rmse showing that this balancing item is difficult to forecast (see Fig. 4.3). Aggregation of industry forecasts to give an overall EPO forecast is on average slightly better than the forecast of the total. There were five out of six origins where the aggregate forecast was more accurate than the direct forecast. It is worth remembering that industry specific forecasts have value, in terms of matching skills to future workload, even if aggregation does not lead to greater overall accuracy.

Table 4.7. Summary accuracy of ARIMA forecasts of EPO annual filings forecasts by Industry

		Industry										
		A	B	C	D	E	F	G	H	other	Aggregated	Total EPO
	Std Error	347	660	720	118	173	353	528	428	91		3211
Accuracy by origin (rmse)	1	256	197	582	2	47	119	846	404	324	1640	1129
	2	554	711	553	51	8	167	529	1834	56	4540	3754
	3	1224	905	1697	183	320	546	979	1288	91	7050	9750
	4	505	1190	651	44	241	988	350	1056	1166	4189	2352
	5	61	71	272	121	177	606	930	189	439	14	1839
	6	339	179	1330	152	41	609	3646	2262	2938	11498	10593
Overall rmse		612	685	982	113	181	585	1645	1382	1310	6093	6212
Overall mape		3.13	2.98	4.31	4.89	3.90	5.83	5.80	6.14	168.7	4.32	4.38

A: human necessities, B: performing operations/transport, C: chemistry; metallurgy, D: textiles; paper, E: fixed constructions, F: mechanical engineering, G: physics, H: electricity.

For this section, results for blocs are summarised in Table 4.8, results for industries are summarised in Table 4.9 and results for total EPO filings are summarised in Table 4.10. In order to judge comparative accuracy, the rank of each method for overall rmse is shown.

The lower the rank, the more accurate is the forecasting method. In addition, in order to gain an idea of how important the differences in rank are, a geometric mean is given. The geometric mean for approach i is defined below, where n filings series have been forecast:

$$GM_i = \sqrt[n]{\prod_{j=1}^{n} \frac{RMSE \text{ for series } j \text{ for approach } i}{\min(RMSE) \text{ for series } j}}$$

The reason for this measure is that it shows an average proportion indicating how far an approach deviates, on average, from the most accurate result.

4.2 ARIMA univariate (monthly data)

Aggregating monthly forecasts over time to provide annual forecasts does not provide greater accuracy, compared to forecasts based on annual data, for any bloc (see Table 4.8). Although aggregation across blocs gives a more accurate forecast of total EPO filings than the direct forecast of total

EPO filings; these are both less accurate than their forecasts based on annual data (see Table 4.10). In Fig. 4.4, the greater accuracy of the annual ARIMA forecast of total EPO filings is visible. Aggregation over the monthly forecasts leads to greater accuracy than the annual industry models for only three out of the nine industries (see Table 4.9). Aggregation of these industry forecasts gives a less accurate forecast of total EPO filings than the aggregated direct annual forecast of total EPO filings (see Table 4.10).

Table 4.8. Overall rmse levels for blocs (Rank of method shown in *italics*)

rmse	EPC		JP		US		Rest of World		Average rank	GM
Univariate ARIMA using annual data	2660	*5*	1186	*5*	1699	*7*	1638	*2*	4.8	1.29
Univariate ARIMA using monthly data	2721	*7*	1353	*6*	1675	*6*	1709	*6*	6.3	1.35
VAR using annual data	5226	*8*	2081	*8*	4935	*8*	1873	*7*	7.8	2.37
VAR using monthly data	1684	*1*	1866	*7*	1243	*2*	2095	*8*	4.5	1.27
Univariate DLM using annual data	2674	*6*	1077	*2*	1581	*5*	1647	*3*	4.0	1.24
Univariate DLM using monthly data	1898	*3*	1084	*3*	1380	*3*	1659	*4*	3.3	1.10
Multivariate DLM using annual data	2346	*4*	1152	*4*	1575	*4*	1681	*5*	4.3	1.23
Multivariate DLM using monthly data	1811	*2*	986	*1*	1169	*1*	1636	*1*	1.3	1.02

rmse defined in Table 4.5.

4.3 ARIMA multivariate (annual data)

For VAR modelling, the order of the process p is normally determined by likelihood ratio tests or by information criteria. However, in the case of yearly data, shortage of observations dictates the maximum value of p. For the four ($k = 4$) EPO bloc series, there are fifteen years ($T = 15$) of differenced data. The constraint on the number of degrees of freedom is:

$$(T-p) - kp > 0, \text{ thus } p < T/(k+1).$$

This means that the maximum value for p is two for blocs. For modelling individual industries (where $k = 9$) the situation is worse with a maximum value for p of one.

For blocs, the forecast accuracy of the VAR model over the last available six years is poor, and it is uniformly worse for each country than the univariate result (see Table 4.8). The shortage of degrees of freedom may have contributed to this poor performance. Repeating the exercise for industries (see Table 4.9) results in a poor forecasting performance by the VAR models. The shortage of degrees of freedom is more severe here as there are more industries than blocs.

4.4 ARIMA multivariate (monthly data)

In order to use stationary data, the data were transformed to $(1-B)(1-B^{12})X_t$. The number of lagged terms to include in the VAR was decided by Bayesian Information Criterion (BIC). Although the maximum value for the number of lags was of the order of twenty or so, the criterion led to the choice of low values. For the blocs, the number of lagged terms was 2, for industries, the number of lagged terms was 1. Note that the information criterion indicates that recent annual changes are useful in prediction, i.e. lags of 1 or 2 means that the data used for the forecasts are between one and two years old.

In addition, it should be noted that the forecasting of monthly data (in VAR) is more problematic than for annual data. Since no data later than the current origin are available for forecasting, the lagged data required in the forecast have to be replaced by forecasts. This means that monthly forecasts for all countries (or industries) are made recursively one month at a time.

The accuracy per bloc is similar to univariate ARIMA (monthly), where there are improvements for the EPC and the US (see Table 4.8). For the overall EPO filings, the forecast based on the aggregated bloc forecasts yields a rmse of 5 307 (see Table 4.10). Similar results for the industries within the EPO are shown in Table 4.9. Industry by industry, the monthly VAR accuracy is similar to monthly univariate accuracy. Aggregating to forecast annual filings, the monthly VAR is more accurate than the corresponding univariate figure for four out of nine industries (industries C, F, G and H). For the total EPO filings, there is an increase in accuracy with a rmse of 4 462 compared to 6 743 for aggregated monthly ARIMA (see Table 4.10).

4.5 DLM univariate (annual data)

Forecasts were computed recursively, using the same initial estimation region as defined in Table 4.4. The accuracy of forecasts using bloc annual data is summarised in Table 4.8. The accuracy of the dynamic linear model for the industry groupings is summarised in Table 4.9. Broadly, the DLM forecasts are similar to the comparable ARIMA ones.

4.6 DLM univariate (monthly data)

As explained earlier, the dynamic linear model was used on monthly data with stochastic trigonometric seasonal factors. Here, for blocs, the monthly rmse values are consistently lower than the comparable ARIMA rmse values. Aggregating across blocs to get an overall EPO filings forecast was more accurate than the direct univariate forecast for both aggregated monthly data and annual (see Table 4.10). The accuracy of forecasts by industry is summarised in Table 4.9. For six out of nine industries, the DLM using monthly data was more accurate than the annual DLM. We see there is no additional accuracy to be gained here by aggregating across industries (see Table 4.10). In Fig. 4.4, we can see that this forecast of total EPO filings tends to be more accurate than the annual DLM and the ARIMA models.

4.7 DLM multivariate (annual data)

The model chosen for forecasting was the homogeneous model that constrains the elements of the covariance matrix driving the disturbances to be proportional to one another. There is a halfway house of *trend homogeneity*, where the constraint of proportionality only applies to level and slope disturbances, not the error term. Experimentation showed that full homogeneity led to more accurate forecasts, so only these are reported. The accuracy of the multivariate dynamic linear model using annual data for EPO blocs is summarised in Table 4.8 and for industries in Table 4.9. The multivariate DLM is of similar accuracy to the univariate DLM.

4.8 DLM multivariate (monthly data)

For monthly data, the accuracy is summarised in Table 4.8 for EPO blocs and Table 4.9 for EPO industries. For both blocs and industries this model has a similar level of accuracy to the univariate DLM.

Table 4.9. Overall rmse levels for industries (IPC main fields of technology). Ranks in *italics*

	A		B		C		D		E		F		G		H		other		Average rank	GM
Univariate ARIMA using annual data	612	*1*	685	*4*	982	*4*	113	*4*	181	*1*	585	*5*	1645	*6*	1382	*6*	1310	*7*	4.2	1.28
Univariate ARIMA using monthly data	907	*5*	626	*3*	1252	*6*	119	*5*	390	*6*	537	*4*	1458	*3*	785	*2*	1384	*8*	4.7	1.38
VAR using annual data	1406	*8*	1712	*8*	2303	*8*	231	*7*	527	*8*	967	*8*	1861	*7*	1600	*7*	1284	*4*	7.2	2.27
VAR using monthly data	1243	*7*	781	*6*	1212	*5*	254	*8*	394	*7*	469	*3*	1016	*1*	749	*1*	1289	*5*	4.8	1.48
Univariate DLM using annual data	639	*2*	740	*5*	957	*3*	92	*1*	220	*4*	587	*6*	1630	*5*	1380	*5*	1198	*2*	3.7	1.28
Univariate DLM using monthly data	676	*3*	539	*1*	859	*1*	108	*3*	216	*3*	408	*1*	1457	*2*	809	*3*	1300	*6*	2.6	1.13
Multivariate DLM using annual data	1050	*6*	1061	*7*	1360	*7*	132	*6*	226	*5*	707	*7*	2629	*8*	2397	*8*	1239	*3*	6.3	1.76
Multivariate DLM using monthly data	717	*4*	546	*2*	863	*2*	103	*2*	195	*2*	437	*2*	1623	*4*	1190	*4*	1074	*1*	2.6	1.16

A: human necessities, B: performing operations/transport, C: chemistry; metallurgy, D: textiles; paper, E: fixed constructions, F: mechanical engineering, G: physics, H: electricity. rmse defined in Table 4.5.

Table 4.10. Overall rmse levels for different aggregated forecasts of total EPO filings

Type of Aggregation	Approach	rmse	Rank	GM
	Univariate ARIMA using annual data	6212	13	1.44
As a univariate	Univariate ARIMA using monthly data	7374	17	1.70
time series	Univariate DLM using annual data	6623	16	1.53
	Univariate DLM using monthly data	4478	3	1.03
	Univariate ARIMA using annual data	6286	14	1.45
	Univariate ARIMA using monthly data	6511	15	1.50
	VAR using annual data	7517	18	1.74
As an aggregate	VAR using monthly data	5307	6	1.23
across blocs	Univariate DLM using annual data	5796	9	1.34
	Univariate DLM using monthly data	4328	1	1.00
	Multivariate DLM using annual data	5989	10	1.38
	Multivariate DLM using monthly data	4578	5	1.06
	Univariate ARIMA using annual data	6093	11	1.41
	Univariate ARIMA using monthly data	5691	8	1.31
	VAR using annual data	9501	19	2.20
As an aggregate	VAR using monthly data	4462	2	1.03
across industries	Univariate DLM using annual data	6126	12	1.42
	Univariate DLM using monthly data	4567	4	1.06
	Multivariate DLM using annual data	10098	20	2.33
	Multivariate DLM using monthly data	5417	7	1.25

rmse defined in Table 4.5.

5 Analysis of comparative accuracy

In order to judge whether the observed differences in accuracy matter it is helpful to test the significance of the differences. Here we use the Friedman test for this task. The ranking of each treatment (forecasting method) is calculated across the available blocks (forecasting origin and time series)[3]. The test first evaluates if the differences in the ranks can be explained as random variation between methods of similar accuracy, or if the differences in ranks are due to greater accuracy of one or more methods. If the hypothesis of no difference in accuracy can be rejected, the signifi-

---

[3] Note that *bloc* is a national grouping; *block* is a grouping of data in this hypothesis testing exercise.

cantly more accurate methods can be identified. For details of the Friedman test, see Conover (1999).

We carry out three comparisons, over blocs, over industries and over total EPO filings. For EPO planning purposes, forecasts of filings by industry and of total filings are of particular interest.

In each case, the block data are the errors for a one-year ahead forecast over six origins and over the available series. For example for blocs, there are four time series (EPC, JP, US and rest of the world) and six origins giving twenty-four blocks.

For blocs, the multivariate version of the dynamic linear model using monthly data is most accurate, but it is only significantly better than the multivariate ARIMA using annual data, using the VAR model. In this case, the main result is that the annual VAR is significantly worse than the other methods.

For industries, the univariate monthly DLM is significantly more accurate than all methods except the multivariate monthly DLM. The multivariate annual versions of ARIMA and the DLM are significantly worse than the remaining models.

For the total EPO filings series, the methods include univariate forecasts of the series and forecasts aggregated over blocs or industries. Three of the four most accurate methods use the monthly univariate DLM, either on the total filings directly or aggregating across blocs or across industries. The use of monthly data dominates the rankings by accuracy.

One of the main issues of this study concerns the advantages of a multivariate model over a univariate model. Is there evidence of information contained in the time series of filings for one bloc or industry that will improve forecasting of filings in other blocs or industries? The answer is no; for blocs a multivariate model was most accurate but not significantly more accurate than its univariate counterpart; for industries there was no evidence of greater accuracy from a multivariate model. This conclusion is conditional on the industrial classification used here. It is not definitive for all possible classifications.

For the issue of aggregation over time, the use of monthly data, rather than annual data, does lead to greater accuracy in terms of ranking. In terms of significantly greater accuracy, this is achieved for industries where the most accurate two methods are significantly more accurate than their versions using annual data. (DMU vs DAU and DMM vs DAM). A similar statement applies to forecasts of total EPO filings, (Bloc DMU vs Bloc DAU and Total DMU vs Total DAU).

Table 4.11. Methods in descending order of accuracy by bloc, industry and total EPO filings

For blocs

Framework	Frequency		Average Rank	Is significantly more accurate than:
DLM	Monthly	Multivariate (DMM)	3.6	AAM
ARIMA	Annual	Univariate (AAU)	4.0	AAM
DLM	Monthly	Univariate (DMU)	4.2	AAM
DLM	Annual	Univariate (DAU)	4.2	AAM
ARIMA	Monthly	Multivariate (AMM)	4.4	AAM
ARIMA	Monthly	Univariate (AMU)	4.5	AAM
DLM	Annual	Multivariate (DAM)	4.5	AAM
ARIMA	Annual	Multivariate (AAM)	6.5	

For industries

Framework	Frequency		Average Rank	Is significantly more accurate than:
DLM	Monthly	Univariate (DMU)	3.3	DAU, AMU, AMM, AAU, AAM, DAM
DLM	Monthly	Multivariate (DMM)	3.7	AAM, DAM
DLM	Annual	Univariate (DAU)	4.2	AAM, DAM
ARIMA	Monthly	Univariate (AMU)	4.4	AAM, DAM
ARIMA	Monthly	Multivariate (AMM)	4.5	AAM, DAM
ARIMA	Annual	Univariate (AAU)	4.5	AAM, DAM
ARIMA	Annual	Multivariate (AAM)	5.6	
DLM	Annual	Multivariate (DAM)	5.8	

For EPO total

Framework	Frequency		Average Rank	Is significantly more accurate than
DLM	Monthly	Univariate (Bloc DMU)	5.0	Bloc AMU, Bloc AAU, Ind DAU, Total AMU, Bloc DAM, Bloc DAU, Bloc AAM, Total DAU, Ind AAM, Ind DAM
DLM	Monthly	Univariate (Total DMU)	5.2	Bloc AMU, Bloc AAU, Ind DAU, Total AMU, Bloc DAM, Bloc DAU, Bloc AAM, Total DAU, Ind AAM, Ind DAM

Table 4.11. (cont.)

Framework	Frequency			Average Rank	Is significantly more accurate than
For EPO total					
DLM	Monthly	Multivariate	(Bloc DMM)	6.2	Bloc AAM, Total DAU, Ind AAM, Ind DAM
DLM	Monthly	Univariate	(Ind DMU)	6.2	Bloc AAM, Total DAU, Ind AAM, Ind DAM
ARIMA	Monthly	Multivariate	(Ind AMM)	8.2	Ind AAM, Ind DAM
ARIMA	Monthly	Univariate	(Ind AMU)	8.7	Ind AAM, Ind DAM
ARIMA	Monthly	Multivariate	(Bloc AMM)	9.5	Ind DAM
DLM	Monthly	Multivariate	(Ind DMM)	9.7	Ind DAM
ARIMA	Annual	Univariate	(Total AAU)	10.5	Ind DAM
ARIMA	Annual	Univariate	(Ind AAU)	10.5	Ind DAM
ARIMA	Monthly	Univariate	(Bloc AMU)	11.7	
ARIMA	Annual	Univariate	(Bloc AAU)	11.8	
DLM	Annual	Univariate	(Ind DAU)	11.8	
ARIMA	Monthly	Univariate	(Total AMU)	12.2	
DLM	Annual	Multivariate	(Bloc DAM)	12.2	
DLM	Annual	Univariate	(Bloc DAU)	12.2	
ARIMA	Annual	Multivariate	(Bloc AAM)	12.7	
DLM	Annual	Univariate	(Total DAU)	13.2	
ARIMA	Annual	Multivariate	(Ind AAM)	15.2	
DLM	Annual	Multivariate	(Ind DAM)	17.7	

For the issue of aggregation over blocs or industries leading to more accurate forecasts of total filings, there was no significant evidence of this.

To achieve the greatest accuracy with one forecasting method, the univariate DLM with monthly data is the most appropriate choice.

If by some diktat, the use of monthly data was forbidden, a similar analysis can be carried out just using annual data. For blocs, the VAR model with annual data is significantly less accurate than other methods. For industries, the univariate ARIMA and DLM are significantly more accurate than their multivariate counterparts. For total EPO filings there was no significant difference detected between the methods. To achieve the greatest accuracy using annual data, either univariate ARIMA or univariate DLM could be used.

6 Forecast accuracy over a longer horizon

The analysis has concentrated on accuracy for the one year ahead forecast. One explanation is that accuracy over short horizons is a guide to accuracy over longer horizons. Another is that a wide-ranging comparison of accuracy over so many approaches and so many different series would have become (even more) unwieldy with more than one horizon. At this stage, two possible univariate forecasting strategies have been identified:

- Univariate DLM using monthly data.
- Univariate ARIMA using annual data.

A third method can be added to represent multivariate modelling:

- Multivariate DLM using monthly data

A comparison across horizons up to four years ahead is carried out using these methods.

The accuracy of the two approaches when applied to blocs is summarised by horizon in Table 4.12. Note that the rmses by horizon are calculated using fewer errors as the horizon increases (6 errors for one year ahead, 3 errors for four years ahead). The accuracy of the approaches for industries is summarised in Table 4.13. The accuracy deteriorates with length of horizon for all methods, from a mape of around 5% at one year to around 30% for four years ahead. The one to four year horizon forecasts of total EPO filings by the DLM (monthly) are shown in Fig. 4.5. It can be seen that the deterioration of accuracy with horizon is due to the forecasts with origins of 1994 to 1996. By 1997, the forecasts have adapted to the steeper trend.

Table 4.12. Comparison of the accuracy of the two broad approaches over a 4 year horizon (blocs)

	Approach	Horizon	EP	JP	US	Rest of World	blocs aggregated to give EPO forecast	Total EPO
rmse	ARIMA	1	2424	1174	1467	1629	5799	5751
		2	5018	2068	2448	1898	10086	9976
		3	9520	3252	5057	3028	19806	19137
		4	15325	4881	8512	4518	32588	32494
	DLM (monthly Univ.)	1	1898	1084	1380	1659	4328	4478
		2	4321	1739	2824	2141	8002	8818
		3	7618	1626	5875	3326	14478	14649
		4	12128	2746	7388	4603	24991	26957
	DLM (monthly Multiv.)	1	1811	986	1169	1636	4578	
		2	4007	1543	2431	2468	8678	
		3	7066	1837	4204	4080	14680	
		4	11398	2901	7728	6354	26915	
mape	ARIMA	1	4.61	4.69	3.29	8.48	4.50	4.60
		2	8.66	10.89	6.66	13.34	8.69	8.31
		3	17.44	18.01	13.05	23.03	16.73	16.13
		4	27.02	27.29	20.86	32.36	25.66	25.65
	DLM (monthly Univ.)	1	3.16	6.45	3.62	8.87	3.10	3.15
		2	8.07	10.02	7.64	15.18	6.20	7.24
		3	14.17	8.56	14.09	24.38	12.06	11.62
		4	21.42	15.11	18.07	32.73	19.30	20.54
	DLM (monthly Multiv.)	1	2.91	5.21	3.00	11.37	3.23	
		2	7.28	7.88	6.67	20.18	7.19	
		3	13.06	8.42	10.32	28.77	11.73	
		4	20.01	15.21	18.80	36.38	20.67	

rmse, mape, defined in Table 4.5.

Table 4.13. Comparison of the accuracy of the two broad approaches over a 4 year horizon (industries)

	Approach	Horizon	A	B	C	D	E	F	G	H	other	Industries aggregated to give EPO forecast
rmse	ARIMA	1	525	643	1014	112	188	534	1636	1165	1323	5673
		2	1196	1418	2117	198	460	852	2859	2206	1318	10385
		3	2198	2632	3121	312	811	1472	3817	4302	2217	19186
		4	361	407	5085	468	1187	2294	6694	7213	2449	32419
rmse	DLM (Monthly Univ.)	1	676	539	859	108	216	408	1457	809	1302	4567
		2	1420	1169	1750	193	364	799	2740	2435	814	8747
		3	2387	1921	2534	261	490	1203	4707	3883	2221	15626
		4	3810	2948	3485	376	689	2016	6097	6455	2564	27731
rmse	DLM (Monthly Multiv.)	1	717	546	863	103	195	437	1623	1190	1082	5417
		2	1543	1279	1679	173	328	821	2743	2850	1111	10885
		3	2640	2172	2627	243	496	1291	4004	4947	2221	19005
		4	4197	3413	4160	376	716	2132	6807	8419	2564	32676

Table 4.13. (cont.)

	Approach	Horizon	A	B	C	D	E	F	G	H	other	Industries aggregated to give EPO forecast
mape	ARIMA	1	2.23	3.23	4.93	4.92	4.28	5.94	5.73	5.75	145.27	4.26
		2	5.42	6.74	9.93	7.78	11.32	9.27	11.10	10.33	391.92	8.72
		3	10.76	12.13	13.77	14.46	20.57	16.49	18.14	19.54	683.84	16.24
		4	18.60	19.10	23.18	23.06	29.29	23.74	28.71	30.48	85.27	25.52
mape	DLM (Monthly Univ.)	1	2.50	2.71	3.61	4.78	5.29	3.92	5.41	3.95	93.55	3.30
		2	5.96	5.00	8.38	9.71	8.75	7.52	11.80	11.74	81.45	7.07
		3	10.96	8.12	11.61	13.04	10.77	11.79	19.01	16.91	100.00	12.96
		4	18.42	13.69	15.46	17.98	15.84	20.83	24.65	27.02	100.00	21.58
mape	DLM (Monthly Multiv.)	1	2.79	2.30	3.63	4.69	4.23	4.74	5.40	5.27	77.83	4.03
		2	6.57	5.70	8.32	8.64	6.63	7.74	11.33	12.51	88.45	8.59
		3	12.55	9.74	11.86	11.87	10.70	13.47	18.44	21.42	100.00	15.74
		4	20.87	16.02	18.60	16.94	16.67	22.11	27.56	34.31	100.00	24.98

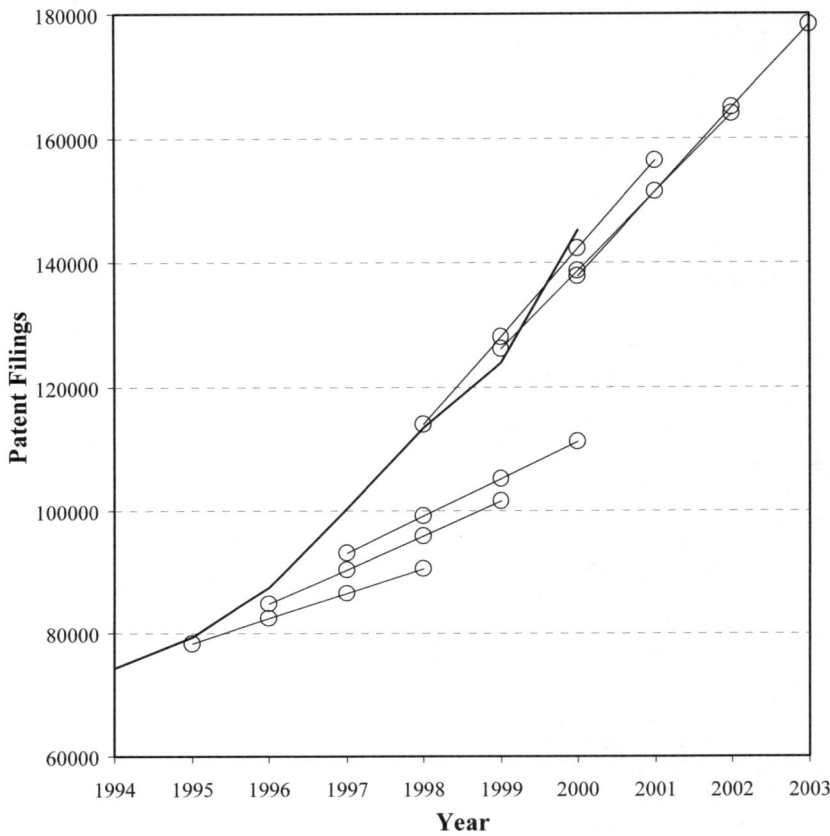

Fig. 4.5. DLM (Monthly) forecasts of total EPO filings with one to four year horizons for origins 1994 to 1999

In Table 4.14, the geometric mean of the rmses of the methods compared to the most accurate is shown for each method. It can be seen that the univariate monthly DLM is more accurate on average than the monthly multivariate DLM, which is in turn more accurate than the annual ARIMA for each of the horizons. In addition the most accurate forecast of total EPO filings is the aggregated blocs forecast by univariate DLM (for all horizons).

Thus the results over horizons up to four years ahead support the findings for a one-year horizon. The DLM approach maintains its superior accuracy over longer horizons.

Table 4.14. Comparison of geometric mean of rmse/minimum(rmse)

Method	Horizon (years)			
	1	2	3	4
ARIMA annual	1.19	1.17	1.25	1.24
DLM monthly univariate	1.06	1.06	1.06	1.01
DLM monthly multivariate	1.08	1.11	1.09	1.12

7 Conclusions

This study has looked at the comparative accuracy of methods of forecasting patent filings at the EPO on four different bases. The first basis for comparison is the issue of whether there is a correlation structure between filings by industry, or bloc, that can be exploited by using a multivariate rather than a univariate forecasting model. For the more accurate models, the monthly DLMs, no significant difference in accuracy was found between comparable models. The implication here is an absence of a useful correlation structure, the series respond similarly to common stimuli, but do not affect each other.

The second basis for comparison is the effect of cross-sectional aggregation across blocs or industries. Again, concentrating on the monthly DLMs, no significant difference in the accuracy of forecasts for total EPO filings was found between the direct forecasts and aggregate forecasts over blocs or industries. This corresponds with Miller's (1998) findings.

The third comparison concerns temporal aggregation, here our findings support those of Rossana and Seater (1995). Forecasts based on monthly data ranked highest in all comparisons. For industries, monthly DLMs were significantly more accurate than their annual counterparts.

The fourth basis of comparison is between the ARIMA and DLM frameworks. For monthly data, the DLM is more accurate, for industries and total EPO filings significantly more so, than monthly ARIMA. For annual data, there was no significant difference between the univariate models. The seasonal modelling of the DLM seems to have better captured the within year variation than the ARIMA model used. This has allowed the DLM to then detect information about changing trends more effectively than the ARIMA model.

It was also demonstrated that the greater accuracy of the univariate monthly DLM persisted over horizons of up to four years.

This univariate forecasting exercise provides a benchmark against which other forecasts drawing on different data sources can be compared. In addition the DLM lends itself to further development by the introduction of possible explanatory variables, such as research and development expenditure, into the state equations.

Chapter 5

Driving forces of patent applications at the European Patent Office: a sectoral approach

Knut Blind

Innovation Systems and Policy, Fraunhofer Institute for Systems and Innovation Research, Karlsruhe, Germany

Modules: C, E
Software used: Eviews

1 Introduction

Patent filings are one of the most important output indicators for the analysis of technological progress. Many studies have analysed the interdependencies between patent applications and the expenditures for research and development (R&D) (Janz et al. 2001, Kortum and Lerner 1999), foreign trade flows (Fagerberg 1988, Greenhalgh 1990, Wakelin 1997) or production (Jungmittag et al. 1999). However, these studies used – in most cases – a product-based approach or the objects of the analyses have been complete countries, because an adequate concordance between technological output and economic sectors (or branches) is not yet available.

Whereas the scope of the above mentioned studies is usually the analysis of economic performance, the aim of this project is to investigate the dependency of patent filings on other economic factors. So the main interest is to uncover and quantify these dependencies. This is done by the following macro approach which is complemented by the micro level analysis reported in Chap. 8.

The consideration of micro-level data alone does not take into account the economic and technological framework of sectors or national economies, which can be assumed to have a certain impact on patent filings. Furthermore, it is much easier to obtain information on sectors than on individual firms, so that it becomes more efficient to get sophisticated forecasts in the future. Most official statistics are provided on the basis of branches and/or national economies, so one solution to this problem is to shift to a higher grade of aggregation and to analyse sectors or national economies. This was done by using a concordance between economic and technological fields.

Time series of the relevant data from 1987 to the present are provided. These data are used to estimate time series models. Due to the restricted length of the time series and the assumption that patent filing depends both on sector characteristics and national idiosyncrasies, the time series are pooled both over sectors, respectively countries, separately and together. The models imply the number of patents as the dependent variable and independent variables like R&D expenditures (source: OECD ANBERD). However, Kortum and Lerner (1999) have observed a change in the patent-R&D relationship in the US, and Janz et al. (2001) have shown, based on German company data, that the relationship between R&D and patenting has become looser. Therefore, other explanatory variables have to be included in the regression analysis. Besides internal sources of information like R&D employees, companies rely in their innovation processes also on externally available knowledge. For science-based industries, publications of the scientific community are important. However, these links are very loose and currently no concordance exists between science classifications and industrial sectors. Therefore, we will rely primarily on the publicly available pool of technological know-how embedded and codified in technical rules and standards. Based on a pooled time series analysis, Blind and Grupp (2000) found a positive relationship between the stock of standards and the output of patent applications. This approach may be extended to other countries.

The growing international interdependencies of companies increase their interfaces with other companies which may also increase the likelihood of patent-related conflicts. An indicator for this networking phenomenon may be the development of relative export flows of sectors (source: OECD STAN). For comparison, we also analyse the influence of economic activity in general measured by the development of value added. Other influential factors to be included may be market structures of sectors like the concentration ratios or the average company size, although data will not be available for all European countries. Nevertheless, all these additional explanatory variables should improve the quality of forecasts of future patent applications at the EPO.

2 Using a concordance between patents and economic factors[1]

There have been a number of attempts in the past to establish a link between technological and economic indicators. However these concordances have not found satisfactory solutions to the following four problems: (1) international comparability, (2) level of disaggregation (3) strong empirical basis, (4) easy applicability to specific problems. Furthermore, since some of these were established, industrial structures have changed, necessitating a change in the nomenclatures.

The earliest attempt at linking technology and industry classifications were done by classifying the patent applications of four countries by NACE classes. This was based on rather an intuitive approach, and was not really based on a systematic analysis that could lead to a well-defined concordance table.

Later, Evenson and Puttnam (1988) use data from the Canadian Patent Office, where patent examiners simultaneously assigned IPC codes, together with an industry of manufacture and sector of use, to each of 300 000 patents granted between 1972 and 1995. On the basis of these data, they established a cross-tabulation between 8 IPC sections and 25 industries, called the Yale-Canada patent flow concordance. The two main problems with this approach which limit its value in terms of practical applications are: (a) it is based on Canadian SIC, which needs to be translated to either ISIC Rev 3 or NACE; and (b) it is not very detailed in terms of IPC codes. An additional difficulty is that the relationship between sectors and technologies has distinctly changed during the period 1972 to 1995.

Verspagen et al. (1994) suggested a concordance scheme between four-digit level IPC subclasses and 22 (2 and 3 digit) industrial classes based on ISIC (rev. 2), the so-called MERIT Concordance. The linkage was established by an intellectual approach, and was based on a similar concordance of Statistics Finland. In this approach, many of the 625 IPC subclasses are linked with different weights to different sectors, so that it is quite time-consuming to calculate statistics for specific sectors.

In the 1980s, the USPTO established a detailed concordance between subclasses of the USPC and 41 unique classes of the US Standard Industrial classification, and this is used to produce regular statistics of US patents by SIC sectors. This is simply done on the basis of examining the definition of each USPC class (and sometimes subclass) and assigning them to one or more of the 41 industrial classes. For our purposes, this

---

[1] The following parts follow closely Schmoch et al. (2003).

concordance has some of the problems already identified above. It is based on the USPC and not the IPC, limiting its applicability to EPO data. Further the industrial classification used is the US-SIC, which needs to be translated into ISIC for practical use.

Greif and Potkowik (1990) computed statistics of patents by industrial sectors, based on an old German national statistical classification scheme (Wirtschaftszweige, WZ79) which is not compatible with the present NACE or ISIC codes. They assigned WZ codes to a sample of 280 applicants in 1983 at the German Patent Office and analysed their patent activities in terms of IPC codes. Again the validity for present purposes is quite limited.

The most recent attempt at defining a concordance between IPC and ISIC codes is by Johnson (2002). As with the earlier work of Evenson and Putnam (1988, see above), this is based on data from the Canadian Patent Office. For 625 IPC subclasses, Johnson defines probabilities of linkages to about 115 different sectors of manufacture and use. However, this interesting method has several limitations. Firstly, the linkage between IPC codes and sectors is defined by examiners of the Canadian Intellectual Property Office, and is not based on the official industrial class of the company to whom the patent is assigned. This is likely to result in a technology bias. Secondly, the Canadian Office stopped assigning sector codes to patents in the grant year 1995, equivalent to about 1991 in terms of first application (priority). Thus, the concordance is quite old, and there is a high probability that the relationship between technology and sectors has changed since then. Thirdly, the sectors are defined in terms of Canadian SIC codes, and have to be translated into ISIC codes, implying certain inaccuracies due to translation. Fourthly, the concordance is based on the determination of 70 000 probabilities of linkage between IPC and ISIC codes. Therefore it can only be handled with the support of a complex software package, consisting of three separate modules. Moreover as input, the user has to provide search results for all IPC subclasses which requires the access to a comprehensive large-scale patent database. Notwithstanding these limitations, the Johnson concordance represents the most advanced suggestion for linking technologies to industrial sectors. However, its adequacy was never tested by a comparison to economic data.

In our analysis, we rely on a concordance developed by Schmoch et al. (2003), which allowed us to produce time series of European and PCT applications for twelve countries (see Table 5.1).

Table 5.1. List of countries

Abbreviation	Country
BEL	Belgium
DNK	Denmark
FIN	Finland
FRA	France
DEU	Germany
ITA	Italy
JPN	Japan
NLD	Netherlands
ESP	Spain
SWE	Sweden
GBR	United Kingdom
USA	United States of America

This approach of Schmoch et al. starts with the selection of industrial sectors at the 2-digit level of NACE or ISIC, with a finer breakdown of the quantitatively important sectors within chemicals, machinery, and electrical equipment, leading to 44 sectors of manufacture. This level of disaggregation is finer than most statistics on economic data, e.g. foreign trade, value added, or R&D expenditure, as provided by OECD, Eurostat or other authorities. It was chosen to be able to show the main differences between the sub-sectors in chemicals, machinery, and electrical equipment industries. Thus a higher level of aggregation to 23 sectors (see Table 5.2) can be achieved by a simple combination (addition) of sub-sectors. We use only 23 sectors due to restrictions in the availability of international comparable economic data on the level of 44 sectors. Moreover it is possible to transfer the NACE-defined fields directly into ISIC-based sectors.

Industrial sectors are defined by the manufacturing characteristic of products, so that it is possible to associate them to technologies. On this basis, technical experts of Fraunhofer ISI associated each of the 625 subclasses of the IPC to one of the 44 industrial categories mentioned above. The IPC subclasses were linked to one field only, even if multiple linkages to other fields were obvious, by applying the principle of main focus.

Table 5.2. List of 23 sectors

Abbreviation	Sector
FOOD	Food products and beverages
TOB	Tobacco products
TEXT	Textiles
WEAR	Wearing apparel, dressing and dyeing of fur
LEA	Leather, leather products and footwear
WOOD	Wood and products of wood and cork
PAP	Paper and paper products
PRIN	Publishing, printing and reproduction of recorded media
PETR	Coke, refined petroleum products and nuclear fuel
CHEM	Chemicals excluding pharmaceuticals
PHAR	Pharmaceuticals
RUB	Rubber and plastics products
NONM	Other non-metallic mineral products
BASM	Base metals
FABM	Fabricated metal products, except machinery and equipment
MACH	Machinery and equipment, n.e.c.
OFF	Office, accounting and computing machinery
ELEC	Electrical machinery and apparatus, n.e.c.
RDTV	Radio, television and communication equipment
MED	Medical, precision and optical instruments, watches and clocks
MOT	Motor vehicles, trailers and semi-trailers
OTRA	Other transport equipment
FURN	Manufacturing n.e.c.

For associating technologies and industries for single companies, an off-line database of Observatoire des Sciences at des Techniques, Paris (OST) was employed, which contains all the data on European and PCT applications without double counting. The information for each patent includes IPC codes, inventors, and applicants with geographical information. This was supplemented by data from Dun & Bradstreet (D&B)[2] which assisted in classifying each applicant by industry. In the D&B database, the industrial activities of firms are described using the US SIC classification, so that they had to be transferred to NACE codes for the purpose of the current project. Although there is no exact correspondence between SIC and NACE codes, it is possible to establish a good association between the classifications at a high level of aggregation (such as the 44 classes mentioned above).

---

[2] See http://www.dnb.com.

Some companies had more than one sector classification in the D&B database. For this purpose, the patents of these companies were split up and fractionally linked to several sectors in order to reduce the heterogeneity between sectors and technologies. The analysis showed that this approach did not reduce, but rather increased the heterogeneity. The reason for this effect is that the correct assignment of singular patents to a sector is not known, but only the overall of the companies' activities in different sectors and technologies. Eventually, each company was attributed to one sector classification.

If the industrial sectors and the associated technology areas were in exact agreement, only the diagonal elements of a cross-tabulation in a matrix of 44 technological fields and 44 industrial fields would be filled. In the case of complete equivalence between technologies and industries, all applications should appear as diagonal elements. However the results of the empirical analysis show that this is not the case, as there is a substantial number of patents in the non-diagonal fields, because the linkage of an IPC code to a sector is "wrongly" assigned, i.e. the IPC code refers to a product range that is not covered by the industrial sector, the technology field can not be linked to one sector in an unambiguous way, but it is linked to several sectors, or the firms in a sector are active in several technologies, partly because they are large multi-product firms, and partly because the products they produce are multi-technology.

The correspondence has a sound empirical basis, as it does not entirely rely on expert assessment in a technological perspective, but on the patent activities of industrial sectors, determined by a very large sample of more than 3 000 enterprises. Moreover, the application of the concordance to specific examples requires a limited amount of work. Database searches have to be performed for only 44 technological fields, defined by a set of IPC subclasses, whereof the results can be transformed into industrial sectors using a 44x44 concordance matrix. Therefore the searches do not require in-house databases, but can be realised by online databases as well. The transformation does not need special software developments and can be done by standard calculation programs.

The suggested concordance can be used for international comparisons, as it refers to international classifications, namely NACE and ISIC for industrial sectors and IPC for patents. With 44 sector fields, the concordance has a reasonable level of disaggregation. A further differentiation would not be useful, as the economic data for international comparisons are not available in a finer breakdown, and the technical interconnections between the sub-sectors would become too strong. Higher aggregation levels can be achieved by a mere combination of sub-sectors.

A specific advantage of the correspondence is the possibility of analysing industrial structures, for instance, by making comparisons across countries, looking for changes over time, or examining differences between large and small enterprises. For such purposes, the technical definitions are kept invariant, whereas different data sets are used for the empirical construction of structural matrices.

The empirical analyses show that a simple, straightforward definition of industrial sectors by technologies would not be appropriate. The two main reasons are that there is often a strong technological interconnection between different sectors, and secondly that large firms produce a broad spectrum of technologies. This result is primarily reflected in the sometimes low importance of the diagonal elements in the concordance matrix, frequently below 20 per cent. This means that in many cases, other sectors contribute more to patents of a specific field than the related *core* sector itself.

The comparison of distribution of the technical fields and industrial classes for different countries, based on country-specific transfer matrices, show a good correlation in many cases. However, there is a relevant number of technical fields with considerable differences. These cases primarily refer to fields with low absolute numbers of patents and firms and to less technology-intensive fields.

These differences between sectors and countries may be, to a certain extent, due to inaccuracies of the association of technological and industrial fields or of the problem to directly link the technological activities of a firm to secondary industrial classifications. Major reasons are, however, the structural differences between the technological activities of industrial sectors in the different countries. The country comparisons in this report exclusively refer to large countries. The structural differences will become even larger if smaller countries are considered, where the technological activities of a specific sector are often dominated by a few firms. Therefore, the comparison of countries has to be handled with care, at least with regard to specific sectors. The concordance matrix developed in this project represents an international average structure.

In summary, the concordance allowed us to calculate the annual number of patent applications in the period from 1985 to 2001 for the 23 sectors in Table 5.2 from the 12 counties listed in Table 5.1. This was the preliminary step for the econometric calculations of the following sections.

3 Empirical results

Time series analyses are a necessary element for predicting the future development of patent applications. The objective of this working step is the identification of determinants for the development of patent applications on an aggregated level. In order to make use of industry indicators, we rely on a new concordance between technology fields classified by IPC codes and industry sectors grouped according the NACE on a two digit level (Schmoch et al. 2003). Before we discuss the selection of adequate industry indicators, we briefly elaborate a second motivation. Since we are able to compare both the results of country and sector analyses, we gain additional insights for the question, whether we still observe country differences by performing analyses on a sectoral level. If we do not find any country influences, this will have strong impacts on the sampling of companies, because then it would be adequate not to consider country quotas any more.

Regarding the indicators, we have selected the following variables. First and most important are the private R&D expenditures. We use the data from the OECD ANBERD database (edition 2003). The values are in PPP US$. The R&D expenditures are the traditional input indicator of the R&D process and it should correlate with its output, the applications of patents. However, there is a recent debate going on, as to whether in the 1990s the development of patent applications is not any more so closely linked to the trends in R&D (Blind et al. 2003a).

Another argument is that the productivity of the R&D process has risen, due to a larger knowledge pool that researchers can rely on. Although it is rather difficult to assess the pool of knowledge relevant for R&D processes, we use as indicator the stock of technical standards as a kind of codified knowledge available to every company (Blind 2004). The hypothesis is consequently that the number of patent applications should correlate positively with an extension of the stock of technical standards. We use the database PERINORM (edition 2003)[3] for constructing data on the stocks of standards for the countries most active in standardisation, like Germany, France, UK and the Netherlands. For all other European countries, the stock of European standards is relevant, because they have reduced national standardisation activities significantly. For Japan and the US, we have no access to adequate time series.

Whereas both the R&D expenditures and the stock of standards are indicators closely connected to the production of new technological knowledge, we use two further indicators. A third general indicator of economic

---

[3] See http://www.perinorm.com/pol/accueil.php.

activity with a possible influence on patent activities, the development of value added, is checked for its influence. We use the data available in the OECD STAN database (edition 2003). These values are recorded in Euros. Finally, the demand for patents representing the intensity of competition is important for explaining the development of patent applications. However, there are no direct indicators measuring the intensity of competition. One can think about the use of concentration indices, like Herfindahl-index or Gini coefficients, but they are not available on an OECD base. One feasible alternative is the trends of export volumes, which is especially relevant for the applications at the EPO, because companies active in foreign markets have a high inclination to apply for patents there. The export data are also taken from the OECD STAN database (edition 2003).

The time series cover at maximum the period from 1987 until 2001, because the data for the R&D expenditures in the OECD database ANBERD start in the year 1987. In general, we use the logarithms of the data, because then the coefficients are elasticities, which represent the change in percentages of the dependent variable after a percentage change of the independent variable.

3.1 The total model

Based on the data of 23 sectors in the twelve countries, we perform in a first step analyses of a so called "total model", which includes a maximum 276 time series. This "total model" assumes identical relations between countries and sectors. Since these are rather unrealistic assumptions, these analyses represent just a preliminary step of the whole time series analysis. We use the following equations for determining the explanatory power of the four indicator variables R&D expenditure (RD), stock of standards (STD), value added (VALA) and exports (EXP).

$$\text{pat}_{ict} = C_{ic} + a_1.RD_{ic;t-n} + g_{ic}.TREND_{ic} + e_{ict} \quad (1)$$

$$\text{pat}_{ict} = C_{ic} + a_1.STD_{ic;t-n} + g_{ic}.TREND_{ic} + e_{ict} \quad (2)$$

$$\text{pat}_{ict} = C_{ic} + a_1.VALA_{ic;t-n} + g_{ic}.TREND_{ic} + e_{ict} \quad (3)$$

$$\text{pat}_{ict} = C_{ic} + a_1.EXP_{ic;t-n} + g_{ic}.TREND_{ic} + e_{ict} \quad (4)$$

Where i = sector; c = country; t = time; n = time-lag; $e_{ict} = \rho_{eict}\text{-}1 + u_{ict}$ with u_{ict} = error term and ρ_{eict} = autocorrelation coefficient of first order. In general, we assume first order autocorrelation in all regressions.

Table 5.3 summarises the results of the four models. In general, the regressions are very significant and have very high R^2 values due to the inclusion of the autocorrelation term.

We find, for all indicator variables, a significant negative coefficient except for the stocks of standards. Since these results are surprising, we do not interpret those further, but proceed immediately with the analyses based on country data.

Table 5.3. Signs of coefficients of the country models

	R&D	Standards	Value Added	Exports
Total model	-(-4)***	+(-5)***	-(-6)**	-(-5)***

Sign: + or -; Lags: number of years in brackets; * = significance at the 10%-level; ** = significance at the 5% level; *** = significance at the 1%-level.
In the case of Japan we use the differences D instead of the logarithms. Data sources: OST patent database; OECD ANBERD; OECD STAN; PERINORM.

3.2 The country models

After presenting and discussing the results of the total model, as based on the time series of all industries in all countries, we analyse the relationship between the various indicators and the patent applications differentiated by country. In Fig. 5.1 the total patent applications of the twelve countries are displayed. We find in all countries a strong increase in the applications in the time period of 1987 to 2001, especially since the middle of the 1990s. The highest level of growth is Spain with almost a 10 fold increase over the period.

The objective of this step in the analysis is to find out whether there are differences between countries regarding the explanatory power of the four indicator variables. We use the following equations for determining the explanatory power of the four indicator variables: R&D expenditure, stock of standards, value added and exports, with variable definitions as in Eqs. 1 to 4 above. We compute regressions for each of the four models for all of the twelve countries based on the time series of the 23 sectors i.

$$pat_{it} = C_i + a_1.RD_{i;t-n} + g_i.TREND_i + e_{it} \qquad (5)$$

$$pat_{it} = C_i + a_1.STD_{i;t-n} + g_i.TREND_i + e_{it} \qquad (6)$$

$$pat_{it} = C_i + a_1.VALA_{i;t-n} + g_i.TREND_i + e_{it} \qquad (7)$$

$$pat_{it} = C_i + a_1.EXP_{i;t-n} + g_i.TREND_i + e_{it} \qquad (8)$$

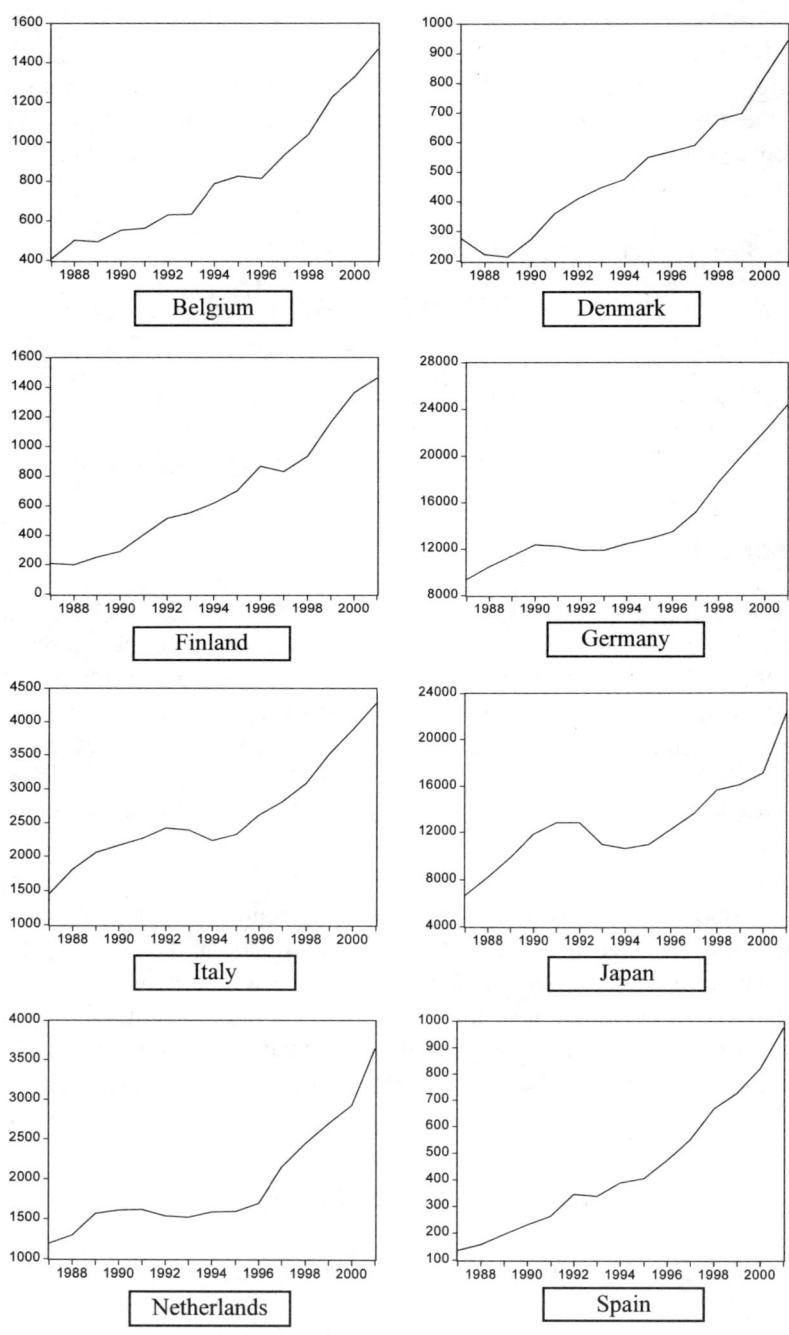

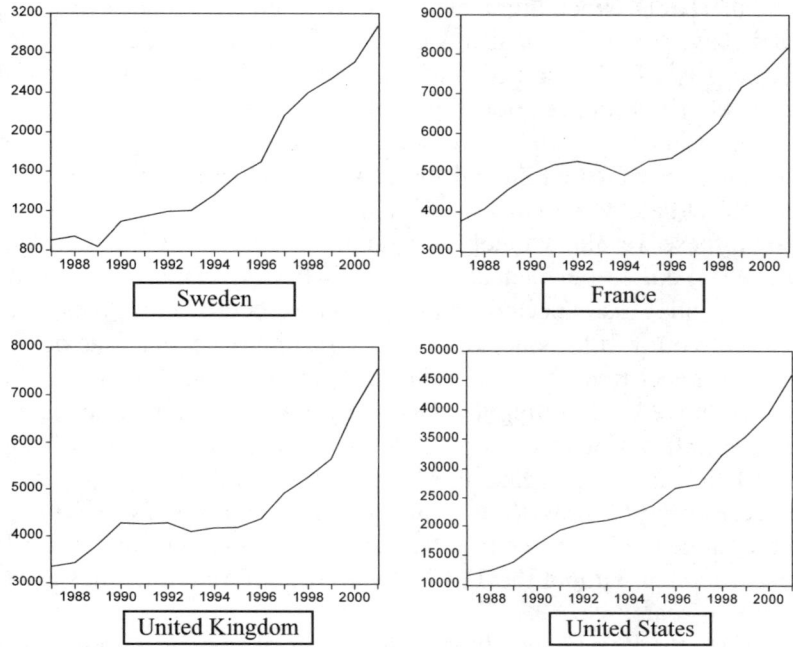

Fig. 5.1. Patent applications by country (Source: OST)

Due to the stable trends in most of the time series, the TREND variable with the basis in 1987 is significant in almost all estimations. The inclusion of the TREND variable is necessary, because the explanatory variables are also characterised by increasing values over time. Furthermore, there are no common deflators, either for R&D expenditures (Hingley 1997, p. 16), value added or for exports. There are also other factors determining patent applications which cannot be operationalised by indicator variables, like strategic patent motives. Consequently, the TREND variable catches all these factors and the exogenous variables are used to explain the deviations of the patent applications from the simple trend extrapolation. Although this approach reduces the likelihood that the exogenous indicator variable is significant, it increases the reliability of the indicator variables in the case of significant coefficients.

First of all, it has to be noted that we find no unique pattern for the explanatory power, either among countries or among the four indicator variables. We report the results of the lagged variable with the highest significance. The input indicator R&D is rather ambivalent. In one quarter of the countries, the coefficient of the R&D indicator is not significant. Only in three small countries, Belgium, Spain and Sweden, do we find the expected positive coefficients of the R&D expenditures. However, the coefficients of the R&D expenditure of the large countries Japan, Germany, France and Italy, and of the small R&D intensive country Finland, are significantly negative. In the industries of these countries, the phenomenon of a relaxed R&D-patenting relationship explained by an increased strategic patenting (cf. Blind et al. 2003a) is obviously more widespread. The sector related analysis promises additional insights because of the strong sector-bias of the R&D-patent relationship.

Although these results cannot be compared directly with those of Hingley (1997), due to our inclusion of the TREND variable, it has to be noted that he finds – as expected – mostly positive coefficients. This is no contradiction to our results, since our R&D expenditures explain the deviations from the linear trend of the variables. If we use another specification of the model above but omitting the TREND variable, then we find for six countries a significant positive impact and only for Germany a significant negative relationship. Regarding the length of the lags, Hingley (1997) finds on average just below three years for the data since 1980. This is comparable to the two years we find for the positive relationships. Regarding the specification without the TREND variable, we find time lags of either one or two years.

In contrast to the R&D expenditures, the stocks of standards are in most cases significant. However, again we find in half of the countries, like Belgium, Denmark, Italy, and the Netherlands, positive and in the other half, like Finland, France, Spain, Sweden, and UK, negative relationships. Although we use the stocks of European standards for the small countries, except the Netherlands, we find significant differences between Belgium, Denmark and Italy on the one hand and Finland, Spain and Sweden on the other hand. It has to be noted that, in the case of positive relationships the lags are mostly five years, whereas in the negative relationships we find mostly shorter lags between one to three years. The positive impacts obviously take longer to influence the patent applications than the negative influences do. Even more than in the case of the R&D expenditures, we expect more insights from the sectoral analyses because the economic impact of standards differs significantly between sectors.

The development of value added is, in the same way as the two other indicator variables, rather ambivalent. For six countries, Denmark, France, Italy, Sweden, UK and US, we find a positive impact on the development of the patent applications. For five countries, Belgium, Finland, Germany, Japan, and the Netherlands, we observe a negative relationship. The lags are evenly distributed between one and six years. Since country differences *per se* should not exist and can again only be explained by different sector distributions or developments, we have to focus once again on the sectoral analyses.

Finally, the export variable has to be discussed. In contrast to the other three indicator variables, for the majority of the countries we find a positive and mostly immediate relationship to the patent applications, including Denmark, France, Italy, Japan, the Netherlands, and Sweden. For the four countries, Finland, Spain, the UK and US we find negative links. In the specification of the model equations without the TREND variable, the export volumes have in ten countries a significant positive impact on the patent applications, with time lags of mostly one up to five years.

Although one could argue that some countries, like Japan or Germany but also Finland or the Netherlands, are very active in exporting goods and are therefore more active in applying for European patent protection, we fall back very quickly to a sectoral argument, because export activities are concentrated in some specific sectors where the competitive advantage is significant. Therefore, in same way as in the discussion of the three other indicator variables, we have to argue and analyse on a sectoral basis and ask whether, in sectors with globalised markets, the patent applications are pushed by the development of export volumes.

Table 5.4. Signs of coefficients of the country models

Coefficient	R&D	Standards	Value Added	Exports	Total
not significant	4	1	1	2	8
Positive	3	4	6	6	19
Negative	5	5	5	4	19

Data sources: OST patent database; OECD ANBERD; OECD STAN; PERINORM.

Table 5.5. Conformity of the signs of significant coefficients of the country models

	R&D	Standards	Value Added	Exports	
R&D		3	4	3	Same
Standards	3		3	6	signs
Value Added	3	5		6	
Exports	3	2	3		
	Different signs				

Data sources: OST patent database; OECD ANBERD; OECD STAN; PERINORM.

Table 5.6. Sign of coefficients, lag structure and degree of significance per country model

Countries	R&D	Standards	Value Added	Exports
Belgium	+(-2)***	+(-5)**	-(-1)*	+(-2)
Denmark	+(-4)	+(-6)**	+(-4)*	+(-3)***
Finland	-(-1)*	-(-3)*	-(-2)***	-(-4)**
France	-(-1)*	-(-1)***	+(-1)***	+(-1)***
Germany	-(-2)***	+(-5)	-(-6)***	+(-3)
Italy	-(-2)***	+(-5)**	+(-5)***	+(-4)**
Japan	-(-3)***(D)		-(-3)***(D)	+(-1)*** (D)
Netherlands	+(-2)	+(-5)**	-(-4)*	+(-1)***
Spain	+(-2)***	-(-2)*	+(-1)	-(-5)**
Sweden	+(-2)**	-(-6)*	+(-2)***	+(-1)**
United Kingdom	+(-1)	-(-1)***	+(-4)***	-(-1)***
United States	+(-4)		+(-3)***	-(-2)**

Sign: + or -; Lags: number of years in brackets; * = significance at the 10%-level; ** = significance at the 5% level; *** = significance at the 1%-level.
In the case of Japan we use the differences D instead of the logarithms. Data sources: OST patent database; OECD ANBERD; OECD STAN; PERINORM.

Finally, we briefly discuss the relationship between the different indicators in the same country. Again the signs of the indicators for each country do not correlate well, although we find a more positive correlation between the coefficients of the export and the value added respective standard equations. On the other hand, the coefficients of the development of the stock of standards and those of value added contradict each other in most cases. Again, we have to look whether we find on the sectoral level similar contradictions.

3.3 The sector models

Since the analyses of the data based on the country models elucidated several puzzling results, we proceed with an analogous analysis of the 23 sectors, based on the data pooled over the twelve countries, i.e. we calculate 23 regressions for each sector based on the available data of the twelve countries c. We use again the four indicator variables in order to explain the development of the patent applications, with variable definitions as in Eqs. 1 to 4 above.

$$\text{pat}_{ct} = C_c + a_1.RD_{c;t-n} + g_c.TREND_c + e_{ct} \qquad (9)$$

$$\text{pat}_{ct} = C_c + a_1.STD_{c;t-n} + g_c.TREND_c + e_{ct} \qquad (10)$$

$$\text{pat}_{ct} = C_c + a_1.VALA_{c;t-n} + g_c.TREND_c + e_{ct} \qquad (11)$$

$$\text{pat}_{ct} = C_c + a_1.EXP_{c;t-n} + g_c.TREND_c + e_{ct} \qquad (12)$$

Tables 5.7 to 5.11 summarise the results of the 23 sectoral analyses. For information, we also computed a model for manufacturing as a total. At first glance, we see no general pattern regarding the signs of the coefficients of the four indicator variables. Although we will also interpret – if appropriate – the results on a sectoral basis, we have divided the sectors in two groups: low-technology and high-technology sectors. Based on Grupp et al. (2000), we define the petroleum, the chemical, the pharmaceutical, the machinery, the office machinery, the electro technology, the radio and television, the medical technology, the motor vehicle and the other transport sector as high-technology sectors. The remaining 13 sectors are defined instead as relatively low-technology sectors. This separation helps us to find more convincing patterns.

Regarding the R&D expenditures, we observe again a rather mixed picture both on the basis of all sectors and after the separation of low- and high-technology sectors, because in one third of the cases we find no significant results at all, in another third we find the expected positive relationship and in the last third a negative link. Even the differentiation into low- and high-technology sectors does not provide significantly more insights. However, the broad majority of high-technology sectors, with the exception of office machinery and medical technology, is obviously not characterised by a positive R&D-patent link, while in the low-technology sectors at least half of the sectors show the expected positive connection.[4] The lags range between one and five years, whereas in the case of negative relations lags of four and five years dominate.

---

[4] If we use the model specification without the TREND variable, we find even less significant connections, but, especially in the low-technology sectors, we mostly find the expected positive relationship between R&D expenditures and patent applications.

Table 5.7. : Signs of coefficients of all sectoral models

	R&D	Standards	Value Added	Exports	Total
not significant	8	8	5	3	24
positive	7	10	7	11	35
negative	8	4	11	9	32

Data sources: OST patent database; OECD ANBERD; OECD STAN; PERINORM

Table 5.8. Signs of coefficients of the low-technology sectoral models

	R&D	Standards	Value Added	Exports
not significant	3	4	1	3
positive	5	5	1	3
negative	5	3	11	7

Data sources: OST patent database; OECD ANBERD; OECD STAN; PERINORM

Table 5.9. Signs of coefficients of the high-technology sectoral models

	R&D	Standards	Value Added	Exports
not significant	5	4	4	0
positive	2	5	6	8
negative	3	1	0	2

Data sources: OST patent database; OECD ANBERD; OECD STAN; PERINORM

Table 5.10. Conformity of the signs of significant coefficients of the sector models

	R&D	Standards	Value Added	Exports	
R&D		6	6	5	Same signs
Standards	4		7	7	
Value Added	6	4		13	
Exports	9	6	2		
Different signs					

Data sources: OST patent database; OECD ANBERD; OECD STAN; PERINORM

The analysis of the influence of the stock of standards elucidates a clearer picture. Although for eight sectors no significant relationship can be detected, in more than double of the sectors a positive connection reveals. The separation into high- and low technology sectors makes the picture even more obvious. The development of the stocks of standards as indicator for the pool of codified technological knowledge has obviously a positive impact on the patent applications in the high-technology sectors, with the exception of office machinery. Among the low-technology sectors, we find three sectors with a negative relationship of the stock of standards to the number of patent applications: the food, the wood and the paper sector. In contrast, five sectors have a positive relation between the respective stocks of standards and the patent applications: the tobacco, the textile, the wearing and dressing, the non-metallic products, and the base metals sectors. The lags cover equally an interval between one and four years.

The results of the estimations of the patent applications based on the time series of the value added are in contrast to those based on the stock of standards. Firstly, in almost half of the sectors, we observe a negative relationship between the value added and the patent applications. Especially in all low-technology sectors except the base metals sector, we find a negative connection. How do we have to interpret this phenomenon? These low-technology sectors are often characterised by relative low growth rates or even by stagnation. On the other hand, the number of patents grows steadily at a rate that is often even apparently exponential. Consequently, we find a negative correlation. The economic activity in these sectors is more or less disconnected with the respective patent activities. In contrast, the dynamic development of high-technology sectors goes in line with the very strong increases in the patenting activities. Taking these observations together, one has to conclude that the development of the economic activity has no influence on the trends in patenting in the low-technology industries, whereas in the high-technology sectors at least a parallel development can be observed, but a direct causality cannot be assumed. Regarding the lags, we find in the case of positive coefficients mostly lags between one and three years. If we observe negative links the lags range between two and six years.

Finally, we have to discuss the results of the estimations with the export volumes as explanatory variable. Whereas we find for the all sectors together that there is no clear picture regarding the sign of the coefficient of the export variable, the separation into low- and high-technology sectors reveals an obvious pattern. In the low-technology sectors, the export volumes correlate negatively with the patent applications in the same way as the development of value added. Only in the food, the leather and the base metals industries do we find a positive correlation. In contrast to this result, all high-technology sectors are characterised by a strong positive influence of the export volumes on the patent applications, with the exception of the special cases of the petroleum sector due to the role of oil and of the other transport industries dominated by the large and heavily integrated aircraft industry. Besides the development of the economic activity in general, the export activities obviously have a strong influence on patent applications. If we consider the lag structures, lags of just one year dominate among the cases of positive correlations, whereas the lags are two to six years in case of negative links, which is similar to the pattern of the estimations based on value added.

Table 5.11. Sign of coefficients, lag structure and degree of significance per sector model[5]

Sectors	R&D	Standards	Value Added	Exports
FOOD	-(-4)**	-(-1)*	-(-3)***	+(-3)**
TOB	+(-3)**	+(-1)***	-(-3)*	-(-3)*
TEXT	+(-4)*	+(-4)**	-(-6)*	-(-5)
WEAR	+(-4)***	+(-1)**	-(-2)***	-(-3)*
LEA	+(-3)	+(-1)	-(-4)***	+(-1)**
WOOD	+(-2)**	-(-1)*	-(-2)	-(-3)*
PAP	-(-4)**	-(-2)**	-(-4)**	-(-2)**
PRIN	-(-4)**		-(-5)**	-(-4)
PETR	-(-5)***	+(-3)***	-(-1)	-(-2)**
CHEM	-(-4)	+(-2)*	+(-3)**	+(-4)**
PHAR	-(-3)	-(-1)	-(-3)	+(-6)***
RUB	-(-5)*	+(-4)	-(-5)**	+(-3)
NONM	+(-2)	+(-4)**	-(-5)**	-(-6)*
BASM	+(-2)	+(-2)***	+(-3)***	+(-3)***
FABM	-(-1)*	-(-4)	-(-4)*	-(-5)**
MACH	+(-1)	+(-2)***	+(-1)**	+(-3)***
OFF	+(-1)*	-(-3)**	+(-4)	+(-1)*
ELEC	-(-4)	-(-2)	+(-2)*	+(-1)*
RDTV	-(-4)**	+(-3)*	+(-1)***	+(-1)***
MED	+(-2)***	+(-1)***	+(-1)**	+(-1)**
MOT	-(-5)**	+(-3)	+(-1)**	+(-1)*
OTRA	-(-3)	+(-2)	+(-2)	-(-5)***
FURN	+(-5)**	+(-2)	-(-3)***	-(-4)**
Total manufacturing	+(-2)***	-(-3)**	+(-2)*	+(-1)**

Sectors defined in Table 5.2. Sign: + or -; Lags: number of years in brackets; * = significance at the 10%-level; ** = significance at the 5% level; *** = significance at the 1%-level. Data sources: OST patent database; OECD ANBERD; OECD STAN; PERINORM.

---

[5] Sign: + or -; Lags: number of years in brackets; * = significance at the 10%-level; ** = significance at the 5% level; *** = significance at the 1%-level.

In the same way as in the analysis of the country models, we finally discuss conformity or contradictions between the signs of the four indicator variables. Most obvious is the strong conformity of the coefficients of the models with value added and export volumes as explanatory variables, because of their strong correlation. For most other combinations we find both identical and different signs. However, it has to be noted that the coefficients of the R&D expenditures contradict each other in two thirds of the cases.

This pattern as well as the same analysis for the country models (in Sect. 3.2) reveals that there is no strong conformity between the direction of influence of the four indicator variables. We observe only a certain correspondence between the signs of the coefficients of value added and export due to their general correlation.

4 Conclusions

We have performed time series analysis based on both twelve country and 23 sector models to estimate the development of patent applications using four indicator variables. At first glance, it is very difficult to detect clear and interpretable patterns. However, the following general conclusions can be drawn based on our results.

The results of the country models are rather heterogeneous and it is difficult to find reasonable explanations for this. However, it has to be noted that the R&D expenditures have only for three countries the expected positive explanatory power for the development of the patent applications. The upsurge in patenting can obviously not be explained by the expansion of R&D expenditures. More powerful for the explanation of the international patent applications on the country basis is the development of the export volumes, because in half of the countries we find a significant positive relationship. In general, the patent activities of most countries we have analysed are biased by a few sectors, which calls for an analysis on a sectoral base.

However, the results of the sector-based analyses are at first glance also not very convincing and we find a similar ambivalence as among the results of the country models. The puzzles of the results can be partly resolved by separating the sectors into low- and high-technology sectors. Firstly, the positive R&D-patent relationship can still be observed in some of the low-technology sectors, whereas it is almost non-existent among the high-technology sectors. Secondly, the stocks of standards have obviously a strong positive influence on half of the high-technology sectors, whereas this is not so clear-cut for the low-technology sectors. Thirdly, value added, but even more the export volumes, have a significant influence on the patent applications. Although we find for most of the low-technology sectors with small or even negative growth rates a negative relationship to the mostly steadily growing patent applications, the picture for the high-technology sectors is a different one. Here, both the development of value added and of the export volumes have a strong and positive influence on the patent applications.

Finally, if we connect the results of the sectoral analyses with the country results, we have to conclude that some few dominating sectors are decisive for the results of the country models. For example, countries like Spain, that are still dominated by low-technology sectors, have a higher likelihood that the R&D expenditures will have a positive influence on their patent applications. On the other hand, the models for countries like Finland, that are dominated by one high-technology sector, will show no significant or even a negative relationship between R&D expenditures and patent applications.

Based on these general results, it seems that future research at the country level should take the sector distributions into account. Furthermore, the application of weighted least squares may be adequate both in the country and sector models in order to account for cross-equation heteroscedasticity. Finally, more comprehensive multivariate models should be developed, which requires further analysis of the interactions between the four indicator variables that have been studied here and others as well.

Chapter 6

Time series methods to forecast patent filings

Gerhard Dikta

Department of Applied Science and Technology, Aachen University of Applied Sciences, Germany

Modules: C, D, E
Software used: S-PLUS with Finmetrics

1 Introduction

This article summarizes some results obtained in the research project "Time series methods to forecast patent filings" which was supported by the research programme "Improvements to methods for forecasting patent filings".

There are basically two methods to forecast Euro-Total filings. The first approach is based on econometric modelling, where the relation of filings to other factors is first formulated as an equation. Then the factors of the model have to be measured quantitatively. At this point there are several time series, representing the quantitative measurements of the factors and the development of filings. In the last step, the unknown parameters of the model are estimated, where the estimation uses the time series related to the factors and the development of filings. Based on this result, filings are forecasted.

In the second approach, no specific model is given in advance, but only some time series which might be related to filings. Then the series are transformed until all of them fulfil the requirements which are necessary to apply some standard statistical models. If this can be done, one can use these models to fit and forecast the transformed series, which finally leads to a forecast of filings.

While the research programme covered both methodologies, this chapter focuses only on the second approach. Some of the results for longer-term forecasts that were obtained in this subproject will be summarized.

The data sets up to the year 2000 are used for modelling und forecasting purposes, i.e. the approach is based on the view of the year 2000. The development of Euro-Total filings over the period 2001–2004 is used to judge the performance of the predictions.

Within a time series analysis (TSA) it is not possible to predict unexpected events somewhere in the future, i.e. TSA has nothing in common with the oracle at Delphi. Therefore, if such an event takes place in the future which affects filings drastically, then it is not possible to predict filing figures very accurately. During the forecast horizon 2001–2005, several events happened which obviously had a large impact on the development of filings at the EPO.

- Year-2000 computer problem
- Terrorist attack of 11th September 2001
- Collapse of the new-economy in 2001 and the drastic drop at the stock exchange
- 2nd Gulf war in 2003

The first event, the Year-2000 computer problem, disappeared thereafter and had no more effect in 2001 and 2002. A possible result of this could be a decrease in filings for the computer industry which could also affect the number of filings from this technical field. Even though one might have been able to use this information within some models, since it was available at the end of the year 2000, it was not taken into account here. The other three events – the terrorist attack on 11th September 2001, the collapse of the new-economy in 2001, and the 2nd Gulf war – all had a strong negative influence on the development of Euro-Total filings. But these kinds of events are unpredictable in advance and therefore no TSA approach could incorporate their corresponding effects in its forecasts.

However, unpredictable events happened in the past and will also happen in the future. Since such events have a strong impact on the development of Euro-Total filings, one might ask whether the long-term forecasting exercise is useful at all. There are two answers to this question. At first a forecast model derived in a TSA is to a large extend objective, i.e. the forecast is not based on any individual assumptions. Thus, if one agrees about the model that is to be used, then one has to accept the forecast. Secondly, one has to remark that long-term forecasts should also come up with a prediction band, i.e. an area which should contain the real development with a certain high probability. The predicted figure itself should be considered as the likely development under the assumption that the development in the future will continue as in the past. Furthermore, future filings figures can also depart from the forecast but should stay within the prediction bands. With this interpretation, long-term forecasts together with their prediction bands are useful as a guideline for planning. Furthermore, the forecast could and should be modified (within the prediction bands) if additional information or ideas about some unexpected events are available which will influence the filing figures in the future.

2 Data description

The data sets that were used are characterized as follows.

1. Filings at the EPO in the period 1979–2000. Source: The EPASYS database of the EPO.

   a. ETT: Euro-Total filings per year. These filings consist of Euro-Direct filings and Euro-PCT filings in the international phase.
   b. ETTEPC: Euro-Total filings from the EPC bloc. This bloc is defined by 20 EPC contracting states, as of December 2000.
   c. ETTJAP: Euro-Total filings from Japan.
   d. ETTUSA: Euro-Total filings from the US.

2. Approximated first national filings over the period 1960–2000. Source: PRI patent families file which is based on priority records extracted from the publications database DOCDB. First filings are given only approximately, i.e. they are counted if they are referenced in a publication stored in DOCDB. See Hingley and Park (2003) for more details.

   a. FFEPC: First filings of the EPC bloc.
   b. FFUSA: First filings of US.
   c. FFJAP: First filings of Japan.

3. Econometric series over the period 1970–2000. Source: Main Science and Technology Indicators of the OECD.

   a. GDP: Global GDP (index 2000).
   b. BERD: Business Expenditures on R&D.
   c. PPP: Purchasing Power Parity (used here as exchange rate).
   d. GDPUSA: GDP of the US (index 2000) (used here as deflator).
   e. RDEPC: BERD of France, Germany, Italy, Netherlands, and United Kingdom changed to US-Dollar according to the individual PPP, aggregated, and finally deflated w.r.t. GDPUSA.
   f. RDJAP: BERD of Japan, changed to US-Dollar according to PPP, and deflated w.r.t. GDPUSA afterwards.
   g. RDUSA: BERD of US, deflated according to GDPUSA.
   h. RDTOT: Calculated as RDJAP+RDUSA+(RDEPC/0.78). RDEPC is divided by 0.78 since the EPC countries used to define RDEPC account for only 78% of the EPO filings in 2000.

Note that a "g" added to one of the abbreviations listed above indicates the corresponding growth rate. As an example, GDPg denotes the growth rate of global GDP.

The econometric data sets used in the analysis are an extension of those data sets which were published in 2003 in the Main Science and Technology Indicators (OECD 2003).

To measure the influence of R&D expenditures on patent filings, several series can be used. Here it was decided to take BERD since these data can be expected to have a direct impact on patent filings. Nevertheless, one critical remark has to be made about these series. The BERD data sets are not totally homogeneous across countries and some of the series even have missing data for some years. Therefore, the RDTOT series, which is used here to measure the overall R&D expenditures, has to be considered critically, since it aggregates series which might not be absolutely comparable.

3 Description of the models

In this section, some major theoretical statistical aspects and definitions are outlined which are needed to use statistical software packages properly. Details about the theory can by found in the textbooks referenced, e.g. Brockwell and Davis (1986, 1996), Lütkepohl (1993), Shumway and Stoffer (1999), or Wei (1990).

If not explicitly defined differently, it is assumed that the time series (TS) X is given by $X = (X_t)_{t \in \mathbf{Z}}$, where the index set is defined by $\mathbf{Z} = \{0, \pm 1, \pm 2, \ldots\}$, i.e. the set of all integers. Thus capital X denotes a random process defined on some probability space. The observed TS, that is, a realization of X, is denoted by $x = (x_t)_{t \in \mathbf{Z}}$. Furthermore, it is assumed that, for every $t \in \mathbf{Z}$, the second moment exists, i.e., $\mathrm{E}((X_t)^2) < \infty$. If the second moment of X_t exists for every $t \in \mathbf{Z}$, the mean function

$$\mu: \quad \mathbf{Z} \ni t \to \mu(t) \equiv \mu_t = \mathrm{E}(X_t)$$

and the autocovariance function

$$\gamma: \quad \mathbf{Z} \times \mathbf{Z} \ni (r,s) \to \gamma(r,s) = \mathrm{E}((X_r - \mu_r)(X_s - \mu_s))$$

are well-defined. Here $\mathbf{Z} \times \mathbf{Z}$ denotes the integer grid in the plane. In the case of $r = s$, this results in the variance function, that is, $\gamma(r,r) = V(X_r)$.

The autocovariance function together with the variance function define the autocorrelation function (ACF)

$$\rho: \mathbf{Z} \times \mathbf{Z} \ni (r,s) \to \rho(r,s) = \frac{\gamma(r,s)}{\sqrt{\gamma(r,r)\gamma(s,s)}}.$$

The values attained by the ACF are in the interval $[-1,1]$.

To be able to forecast the future values of a TS, some kind of regularity has to exist over time in the development of the TS. Here, we assume that the TS is (weakly) stationary, that is,

$$E(X_t) = \mu_t \equiv \mu = const., \text{ and } \gamma(r,s) = \gamma(r+t, s+t), \text{ for all } r,s,t \in \mathbf{Z}.$$

The first assumption ensures that the TS stays on average at the same level over time, while the second assumption guarantees a time-stable dependency structure, that is, $\gamma(r,s)$ depends only on the difference between r and s rather than on the exact time points. This also guarantees that the variances of the TS are stable over time. For a stationary TS, the autocovariance function can be simplified by:

$$\gamma: \mathbf{Z} \ni t \to \gamma(t) = \gamma(t,0).$$

Thus, the variance function reduces to $\gamma(0)$, in which case the ACF reduces to $\rho(h) = \gamma(h)/\gamma(0)$.

Most econometric TS are not stationary due to some non-constant trend and heteroscedastic error terms. These series have to be properly transformed before they are analyzed. Typically, Box-Cox transformation is used on the raw data, i.e.

$$f_\alpha(x) = \begin{cases} \log(x) & : \alpha = 0 \\ (x^\alpha - 1)/\alpha & : 0 < \alpha \leq 1.5 \end{cases}$$

to handle heteroscedasticity, and differencing

$$\nabla X_t = X_t - X_{t-1}, \quad \nabla^d X_t = \nabla(\nabla^{d-1} X_t)$$

to filter out the trend. Thus ∇X_t denotes the difference $X_t - X_{t-1}$, $\nabla^2 X_t$ the difference $\nabla X_t - \nabla X_{t-1}$, etc.

An important stochastic model, which is often used to model the error process, sometimes also called the innovation process, is given by the white noise process $W = (W_t)_{t \in \mathbf{Z}}$. This process is defined by a mean function $\mu_t = 0$ and an autocovariance function γ, with $\gamma(0) = \sigma^2$, and $\gamma(t) = 0$ for $t \neq 0$. Such a white noise process, abbreviated here by $W \sim \text{WN}(0, \sigma^2)$, has no structure and therefore no information about the past observations (except the uncertainty of a prediction) which can be used to forecast the future. In the special case of $W_t \sim N(0,\sigma^2)$, that is, W_t is a centered and normally distributed variable with variance σ^2, the process W is called Gaussian white noise, denoted here by GWN or GWN($0,\sigma^2$). Due to a special property of the normal distribution, a GWN processes $W = (W_t)_{t \in \mathbf{Z}}$ consists of independent and identically normal distributed variables W_t.

The dynamics of a univariate TS X are often modelled according to the Box and Jenkins (1976) approach. Under this approach, it is assumed that the TS, properly transformed, is an autoregressive moving average process (ARMA). For this, let $0 \leq p,q$ be two non negative integers. Then X is said to be an ARMA(p,q) process, if X is stationary and, for every $t \in \mathbf{Z}$,

$$X_t - \phi_1 X_{t-1} - \ldots - \phi_p X_{t-p} = W_t + \theta_1 W_{t-1} + \ldots + \theta_q W_{t-q}, \tag{1}$$

where $W \sim \text{WN}(0,\sigma^2)$. The special cases of ARMA($p,0$) and ARMA($0,q$) are the autoregressive process of order p, denoted by AR(p), and the moving average process of order q, denoted by MA(q), respectively. Equation 1 can be written symbolically in the more compact form

$$\Phi(B)X_t = \Theta(B)W_t, \qquad t \in \mathbf{Z},$$

where Φ and Θ are polynomials of degree p and q, respectively, i.e.

$$\Phi(z) = 1 - \phi_1 z - \ldots - \phi_p z^p, \quad \Theta(z) = 1 + \theta_1 z + \ldots + \theta_q z^q$$

and B denotes the backward shift operator defined by $B^j X_t = X_{t-j}$. Throughout the analysis, it will be assumed that the polynomials Φ and Θ have their roots outside the unit circle. This guarantees that the ARMA process has a unique solution, i.e. the process is invertible and stable.

Closely related to ARMA processes are the autoregressive integrated moving average processes (ARIMA, see Chap. 4). A process X is an ARIMA(p,d,q) process if $\nabla^d X$, the d–times differenced process X, is an ARMA(p,q).

Forecasts based on the ARMA approach incorporate only the history of the corresponding TS. Neither known forthcoming events nor any other kind of information are used. Typically, econometric TS are influenced by other series. In some cases, the series of interest (output) reacts with a certain delay to some other series (input) and one can use the time-lag between the known input and the output to improve the forecasts of the output. This type of analysis is often done under the assumption of a transfer function model (TFM). For this, assume that the input is given by $X = (X_t)_{t \in \mathbf{Z}}$ and the output by $Y = (Y_t)_{t \in \mathbf{Z}}$, such that

$$Y_t = \sum_{i=0}^{\infty} \alpha_i X_{t-i} + \eta_t = \alpha(B) X_t + \eta_t,$$

where the processes X, Y, and η are assumed to be stationary and η is independent of X. Furthermore,

$$\alpha(B) = \frac{\delta(B) B^d}{\omega(B)}, \qquad (2)$$

where the polynomial in the numerator is given by

$$\delta(B) = \delta_0 + \delta_1 B + \delta_2 B^2 + \cdots + \delta_s B^s,$$

and the polynomial in the denominator by

$$\omega(B) = \omega_0 - \omega_1 B - \omega_2 B^2 - \cdots - \omega_r B^r.$$

In addition, if the underlying processes X and η are ARMA processes, i.e.

$$\Phi_x(B) X_t = \Theta_x(B) W_t, \qquad t \in \mathbf{Z}, \qquad (3)$$

and

$$\Phi_\eta(B) \eta_t = \Theta_\eta(B) Z_t, \qquad t \in \mathbf{Z}, \qquad (4)$$

where $W \sim \text{WN}(0, \sigma_X^2)$, $Z \sim \text{WN}(0, \sigma_\eta^2)$, the TFM is also called an ARMAX process. Here it is always assumed that the TFM is ARMAX.

To detect the delay between the input and the output series and the order of the polynomials δ and ω of such a TFM, the cross-correlation function (CCF) is used. For the definition of the CCF, the second moments of X_t and Y_t have to exist for every $t \in \mathbf{Z}$. Denote the corresponding mean functions and autocovariance functions of the processes by $\mu_x(t)$, $\mu_y(t)$, $\gamma_x(t)$, and $\gamma_y(t)$, respectively. The CCF is then defined by

$$\rho_{xy}: \quad \mathbf{Z} \times \mathbf{Z} \ni (r,s) \to \rho_{xy}(r,s) = \frac{\gamma_{xy}(r,s)}{\sqrt{\gamma_x(r,r) \gamma_y(s,s)}},$$

where

$$\gamma_{xy}: \quad \mathbf{Z} \times \mathbf{Z} \ni (r,s) \to \gamma_{xy}(r,s) = E((X_r - \mu_x(r))(Y_s - \mu_y(s)))$$

denotes the cross-covariance function. As the concept of stationarity is basic for the analysis of a univariate process under the Box-Jenkins approach, a similar assumption has to be made in the case of a TFM in order to guarantee that the interrelationship between the two processes stays stable over time. Precisely, one has to assume that the processes are jointly stationary,

$$\rho_{xy}(t+h,t) = \rho_{xy}(h,0) \equiv \rho_{xy}(h)$$

for every $t, h \in \mathbf{Z}$.

The identification procedure of a TFM, where X and Y are jointly stationary, i.e. the determination of $\alpha(B)$ in Eq. (2), is based on the CCF of the prewhitened input series W, see Eq. (3), and the transformed output series defined by

$$\bar{Y}_t = \frac{\Phi_x(B)}{\Theta_x(B)} Y_t,$$

where the polynomials Φ_x and Θ_x are the corresponding ones of the ARMA process X given in Eq. (3). Precisely, one has

$$\alpha_k \propto \rho_{w\bar{y}}(k),$$

for every $k \in \mathbf{Z}$. See Wei (1990) p. 225.

It is worthwhile to consider the idea of prewhitening from a less theoretical point of view. Assume that both the output Y and the input X are driven by another process, say Z. In such a situation, the development of Y and of X is determined by the development of Z. If one would take the CCF of X and Y directly to find the impact of X on Y, one would detect an artificial relationship which is useless for the purpose of forecasting. However, if the input X is white noise, and this is what prewhitening does to the input, there is no structure behind X and therefore it makes sense to assume that now the real impact of X on Y can be observed.

In this chapter, the procedure for identification of a TFM is mainly used to study and detect time-lags between input and output. TFM in general is not used here to model and forecast patent filings, since these models are mainly restricted to a single input series only. To handle more than one input series, a dynamic multivariate regression approach is appropriate. Two types of models will be considered here, which generalize the univariate autoregressive models. In particular, autoregressive distributed lag models, as a direct extension of TFM to more than one exogenous series as input, and vector autoregressive models to model several series simultaneously, will be applied. For this, some of the above given definitions have to be generalized to vector-valued TS.

Assume that $X = (X_t)_{t \in \mathbf{Z}}$ is a p-variate TS, i.e.

$$X_t = (X_{t,1}, X_{t,2}, \cdots, X_{t,p})^T \in \mathbf{R}^p.$$

Then X consists of p univariate TS, $X_1, X_2, \cdots, X_p$, where $X_i = (X_{t,i})_{t \in \mathbf{Z}}$. Such a p-variate TS is called stationary, if the TS X_i and X_j are jointly stationary for all $1 \le i, j \le p$. The expectation μ of such a series is then a vector

$$\mu = (\mu_1, \mu_2, \cdots, \mu_p)^T = (\mathrm{E}(X_{0,1}), \mathrm{E}(X_{0,2}), \cdots, \mathrm{E}(X_{0,p}))^T.$$

Furthermore, the CCF generalizes to the cross-correlation matrix (CCM) for every time-lag k, denoted here by

$$\Upsilon(k) = (\rho_{i,j}(k))_{1 \le i, j \le p},$$

where $\rho_{i,j}(k)$ denotes the CCF at time-lag k of X_i and X_j. If a p-variate process W, with expectation $\mu = 0$, has a CCM Υ, such that $\Upsilon(k) = 0$ for all $k \ne 0$, the process W is called a p-variate white noise, abbreviated by $W \sim \mathrm{WN}(0, \Upsilon)$. In the special case of a p-variate white noise process W which has a multivariate normal distribution, the process will be called a multivariate Gaussian white noise, indicated here by $W \sim \mathrm{GWN}(0, \Upsilon)$. Note that, for $s \ne t$, W_s and W_t are uncorrelated if the process is white noise. Note further, that $W_{t,i}$ can be correlated to $W_{t,j}$.

Consider now a stationary univariate output process Y, a stationary p-variate input process X, and a univariate white noise process W which is uncorrelated to the process X. Assume that the (p+1)-variate process (Y, X) is stationary. Then

$$Y_t = a + \sum_{j=1}^{k} \phi_j Y_{t-j} + \sum_{j=0}^{l_1} \beta_{j,1} X_{t-j,1} + \cdots + \sum_{j=0}^{l_p} \beta_{j,p} X_{t-j,p} + W_t$$

will be called an autoregressive distributed lag model (ADL). This model is an extension of the TFM.

To analyze the interrelation of the single series of a p-variate process X and to obtain simultaneous forecasts of all the univariate TS involved in X, vector autoregressive models are used here. For this, assume that X and W are p-variate stationary processes such that X is stationary and W is white noise. If for every $t \in \mathbf{Z}$,

$$X_t = c + \Phi_1 X_{t-1} + \Phi_2 X_{t-2} + \cdots + \Phi_q X_{t-q} + W_t$$

where $c \in \mathbf{R}^p$ is an intercept and $\Phi_1, \cdots, \Phi_q$ are $p \times p$ transition-matrices, then X is called a vector autoregressive process (VAR) of order q, abbreviated by VAR(q). (See also Chap. 4, Sect. 3). The concept of stability of univariate AR processes carries over to VAR processes correspondingly.

To be precise, a VAR(q) process X with transition matrices $\Phi_1, \cdots, \Phi_q$ is stable, if the eigenvalues of the companion-matrix A, where

$$A = \begin{pmatrix} \Phi_1 & \Phi_2 & \cdots & \Phi_{q-1} & \Phi_q \\ 1_q & 0_q & \cdots & 0_q & 0_q \\ 0_q & 1_q & & 0_q & 0_q \\ \vdots & & \ddots & \vdots & \vdots \\ 0_q & 0_q & \cdots & 1_q & 0_q \end{pmatrix},$$

are in the unit circle. Here 1_q and 0_q denotes the $p \times p$ identity matrix and the $p \times p$ 0 matrix, respectively.

In the application of the VAR approach, one puts together several univariate TS in a vector, which then is a (stable) multivariate TS, to study their relationship statistically. Often this is done in the absence of an econometric model, and therefore statistical tools are necessary for the structural analysis. In the forthcoming application of the VAR approach, Granger-causality, orthogonal impulse response analysis, and orthogonal forecast error variance decomposition will be used for the structural analysis. A detailed description of these statistical tools can be found in the books of Lütkepohl (1993) or in Zivot and Wang (2003). Here only some ideas of the terms listed above are sketched briefly. A univariate TS, Y, Granger causes a univariate TS, Z, if the information or knowledge of Y reduces the mean squared prediction error of the Z series. In an orthogonal impulse response analysis, one studies the reaction of one variable to an external impulse or innovation (or error) of another variable in a system which also includes several other variables. In most cases, all the variables of the system are correlated to one another and therefore a sudden impulse of one variable alone is not realistic since it should be accompanied by an immediate impulse of some of the other variables which are correlated to the modified one without any time-lag. To overcome this difficulty, one can apply an orthogonal impulse response analysis. For this, the system has to be transformed first in such a way that the corresponding error vectors are independent w.r.t. different time points, but consist also of uncorrelated (orthogonal) entries. With such an orthogonal representation a sudden external impulse due to just one variable is realistic and one can then study how an (orthogonal) impulse propagates through the system. For such a study, one needs to have a causal ordering of the variables. In the forthcoming applications, Granger causality is used to obtain such causal orderings. The above described orthogonal representation of the system also allows the variance of the prediction errors to be decomposed w.r.t. the variances of the (orthogonal) innovations (errors), i.e. to apply an orthogonal forecast error variance decomposition (EVD). This method shows the impact and the importance of each variable of the system in the prediction of the variable one is interested in.

In the forthcoming application of the VAR approach, it is obviously desirable to obtain forecasts of patent filings based on assumed knowledge of the development of some other variables of the VAR model. This will be useful when additional information, due to time-lags, can be incorporated into the forecasts in order to obtain so called conditional forecasts, see Zivot and Wang (2003) for further explanation of conditional VAR forecasts.

4 Description of the applied methods and diagnostics

The analysis was carried out with the software S-PLUS including the S+FinMetrics package of the Insightful Corporation (Insightful, 2001, 2002). This package contains a wide variety of powerful tools to model and to analyze time series. A user of this software package has to have at least a basic knowledge of statistics to apply the tools properly.

Every univariate TS was transformed to a stationary one at first. While the finding of a proper transformation of a TS to a stationary one was based on visualization, stationarity was finally checked with the KPSS test, due to Kwiatkowski, Phillips, Schmidt and Shin; see Zivot and Wang (2003), p. 123. In this test, the null hypothesis of a constant trend was checked. Note that a constant trend means that the TS fluctuates around a constant, i.e. it stays stable on average.

The residuals of every model (ARMA, ADL, VAR) were tested to ensure that they were uncorrelated (Ljung-Box test) and normally distributed (Shapiro-Wilks test); see Zivot and Wang (2003), p. 63 and p. 61, respectively.

Stability and invertibility checks of ARMA models, in those cases where prewhitening was applied, were done by obtaining the roots of the corresponding AR and MA polynomials. If they were outside the unit circle, stability and invertitibility was accepted. In the case of VAR models, the eigenvalues of the companion matrix were calculated. Stability of the applied model was accepted if the eigenvalues were within the unit circle.

VAR models were fitted to the data with the S+Finmetrics function VAR. A model was accepted if the corresponding univariate residuals were uncorrelated and normally distributed. Furthermore, the VAR model had to pass a stability check. Since the number of observations of the considered TS are small, only low order VAR models are possible. In most cases, only one VAR model passed the above mentioned tests. The causal ordering, needed for the structural analysis, was obtained by an application of the Wald test; see Zivot and Wang (2003), p. 391. Impulse response analysis and the forecast error variance decomposition were done with the S+Finmetrics functions impRes and fevDec; see Zivot and Wang (2003), p. 393 and p. 399, respectively. Finally, conditional VAR forecasts were calculated through the cpredict function; see Zivot and Wang (2003), p. 389.

ADL models were fitted to the data with the ordinary least square method, see OLS function in Zivot and Wang (2003), p. 171. As for the VAR models, an ADL model was only accepted if the corresponding residuals passed the correlation and the normality test. To obtain the order of the input series, a stepwise reduction technique (SRT) was applied. In particular, a model with a (high) suitable number of lags per TS (input factors) was fitted to the data in the beginning. Then the AIC model selection criterion was calculated (see Lütkepohl 1993, or Zivot and Wang 2003 for a further description of AIC) and the input factor with the parameter whose p-value was the highest in the corresponding t-test (null hypothesis: parameter=0) was taken out of the model. With this modified set of input factors, a new model was fitted and its AIC calculated. If the new model passed the residual diagnostics and, in addition, if it had an AIC less than the model considered before, the new model was taken. Continuing with this stepwise reduction of input factors, the SRT stops finally with a model which passes the residual diagnostics and which has the lowest AIC among those considered.

5 Results of the analysis

5.1 Stationarity

Considering the development of ETT filings over the period 1979–2000, one finds a linear trend from the beginning up to the year 1990, followed by a drop in 1991 and a quadratic or cubic trend until 2000. Thus the ETT series cannot be trend stationary and has to be differenced at least once.

But one-time differencing will filter out only the linear trend of the starting period and not the quadratic or cubic trend of the most recent years. Therefore, at least two-time differencing seems to be appropriate here.

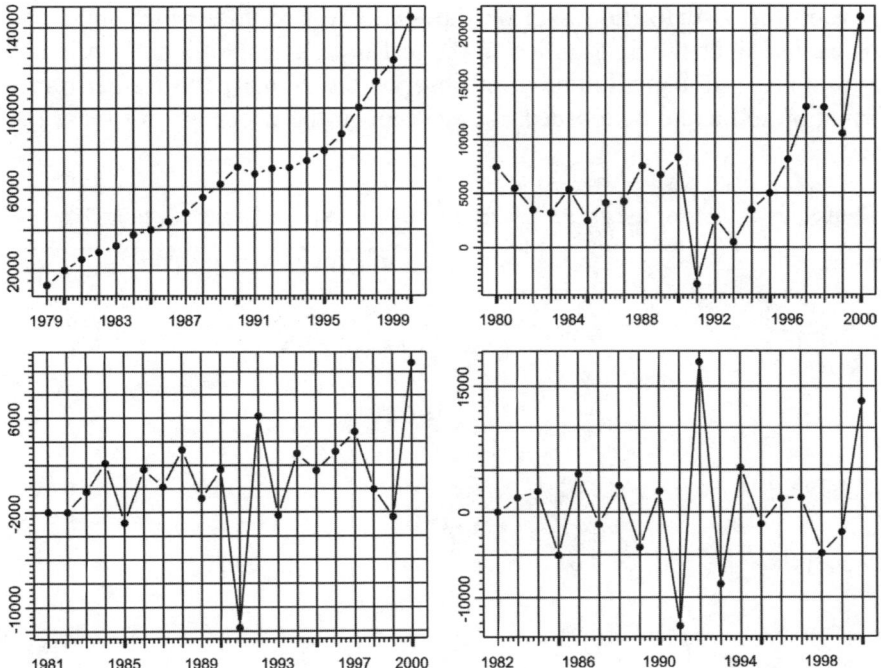

Fig. 6.1. Visualization of ETT (top left), ∇^1 (ETT) (top right), ∇^2 (ETT) (bottom left), and ∇^3 (ETT) (bottom right)

A visualization of the differenced ETT series is given in Fig. 6.1. The linear time trend over the period 1991 – 2000 of the one-time differenced ETT series can be clearly observed, while hardly any time trend is left in the case of two-time differencing (see Fig. 6.1, bottom left). Therefore, based on the visualization, two-time differencing is the first choice to filter out the trend.

However, the KPSS test of the null hypothesis of a constant trend, i.e. when the null hypothesis is that the data fluctuate around a constant (constant trend in S-PLUS notation), shows p-values for one- and two-time differencing of approximately 0.1, while the p-value is far above 0.1 in the case of the three-time differenced ETT series, see Table 6.1. Therefore, based on the p-values of the KPSS test, three-time differencing would be the first choice for a significance level of 10%.

Even though the KPSS test suggests three-time differencing in the cases of the ETT and the ETTUSA and one-time differencing in the cases of the ETTEPC and the ETTJAP series, we decided to use two-time differencing for all the series due to the corresponding visualizations. Here we rely more on the visualization than on the KPSS test results, which might be misleading since the distribution of the test is an asymptotic one while our series consist of a small number of observations. Finally, no further transformation to handle heteroscedasticity seems to be necessary for these series.

Table 6.1. p-values of the KPSS test of the null hypothesis "constant trend"

SERIES	DEGREE OF DIFFERENCING			
	0	1	2	3
ETT	$\ll 0.01$	≈ 0.10	≈ 0.10	$\gg 0.10$
ETTEPC	$\ll 0.01$	$\gg 0.10$	$\gg 0.10$	$\gg 0.10$
ETTUSA	$\ll 0.01$	< 0.05	< 0.10	$\gg 0.10$
ETTJAP	$\ll 0.01$	> 0.10	> 0.10	> 0.10

Similarly, the degree of differencing was chosen for the other series without any further transformations. In Table 6.2 these results are summarized.

Table 6.2. Degree of differencing used for the analysis

SERIES	ETT	ETTEPC	ETTUSA	ETTJAP
DIFFERENCING	2	2	2	2
SERIES		FFEPC	FFUSA	FFJAP
DIFFERENCING		2	2	2
SERIES	RDTOTg	RDEPCg	RDUSAg	RDJAPg
DIFFERENCING	1	1	1	2
SERIES	GDPg		USAGDPg	
DIFFERENCING	1		1	

5.2 Prewhitening and cross correlation

In this section, the influence of research and development (R&D) expenditures on Euro-Total filings is analyzed. Since national R&D expenditures influence the number of national first filings (Hingley 1997), and these first filings are the base of subsequent filings at the EPO, one can expect to find some evidence for such a relationship statistically. Typically, one can expect a time lag between the R&D expenditure and the invention stimulated by it. This time lag together with the actual R&D expenditures can then be used to forecast the future number of patent applications at the EPO.

Based on the transformed series, time lags between series were analyzed according to the prewhitening technique as described above. In the following, this is done as an example for the transformed RDTOTg and ETT series. Both series are transformed according to Table 6.2 and denoted in the following Fig. 6.3 by: xtrans for the input data $\nabla^1 \text{RDTOTg}$; ytrans for the output data $\nabla^2 \text{ETT}$. Note that the input data can be assumed to be a GWN process, since this series passed the Ljung-Box test with a p-value of 0.65 and the Shapiro-Wilks test with a p-value of 0.76. Therefore, the input data show no significant correlation or departure from normality. Thus a further filtering is not necessary in this case.

Considering Fig. 6.2, a significant 5 year delay is observed in the CCF plot (bottom left) before the ytrans output reacts to the xtrans input. This delay gives us the option to use the current value of the input data to predict the output data 5 years into the future.

To visualize this time lag over the total period of observations, Fig. 6.3 shows the development of the pattern of the input data, shifted by 5 years, and the pattern of the output data. Note that only the patterns are comparable here, since the input data had to be multiplied by a constant factor to find a common level for the plot. Obviously, the patterns are similar and the 5 years time lag seems to fit.

In summary, Table 6.3 shows all the results derived by this technique.

It is worthwhile to mention here that the time lag of 6 years, detected between the input process related to RDEPCg and the output process related to ETTEPC, is also verified by the time lag of 5 years found between the processes related to RDEPCg and FFEPC. The latter lag is exactly what one could expect since the national first filings are one year ahead of the corresponding subsequent filings due to the priorities system. But ETTEPC is based to a large extent on subsequent filings.

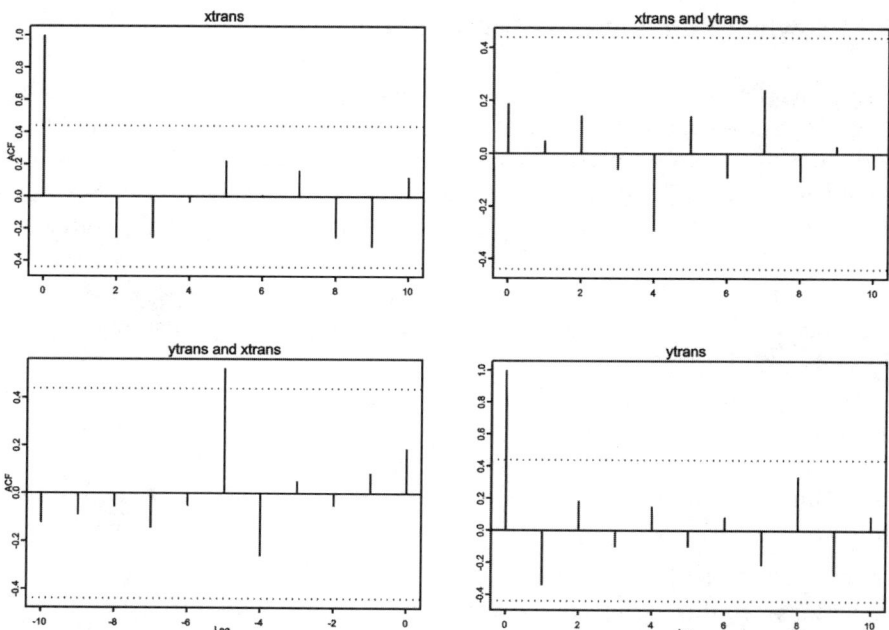

Fig. 6.2. ACF and CCF plots of prewhitened and transformed input xtrans = ∇^1 RDTOTg and output ytrans = ∇^2 ETT

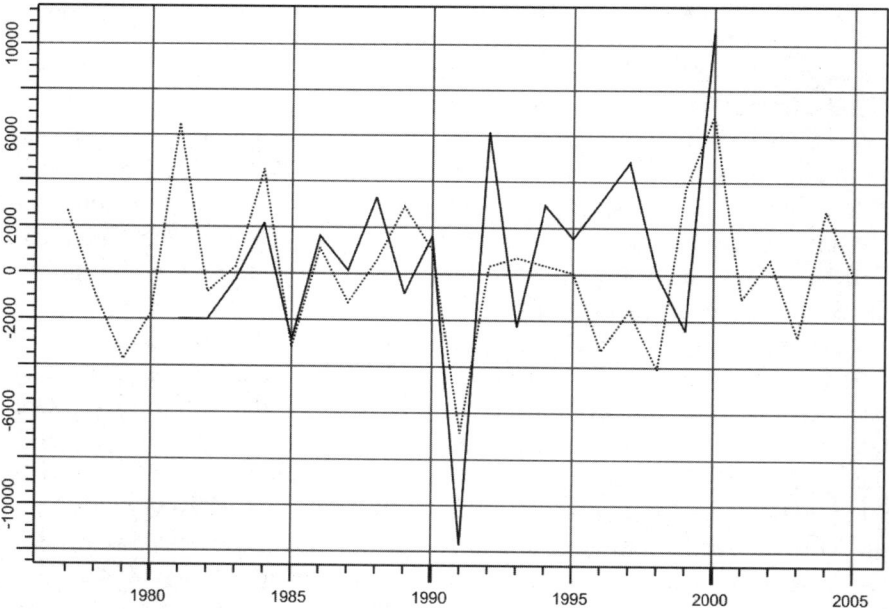

Fig. 6.3. Pattern of input data shifted by 5 years (dotted line, starting in 1977) and pattern of the output process (solid line, starting in 1981)

Table 6.3. Results of the time lag analysis based on prewhitened input and output processes

INPUT				OUTPUT	DELAY sig.
Process	Structure	p-val (L-B test)	p-val (S-W test)	Process	Years
∇^1 RDTOTg	GWN	0.65	0.76	∇^2 ETT	5
∇^1 RDEPCg	AR(1)	0.85	0.97	∇^2 ETTEPC	6
∇^1 RDEPCg	AR(1)	0.85	0.97	∇^2 FFEPC	5
∇^1 RDUSAg	AR(2)	0.76	0.39	∇^2 ETTUSA	5
∇^1 RDUSAg	AR(2)	0.76	0.39	∇^2 FFUSA	1
∇^2 RDJAPg	GWN	0.33	0.97	∇^2 ETTJAP	4
∇^2 RDJAPg	GWN	0.33	0.97	∇^2 FFJAP	-
∇^1 GDPg	GWN	0.63	0.65	∇^2 ETT	0

The processes related to RDJAPg and FFJAP show no significant lag, while a 4 years lag could be detected between the processes related to RDJAPg and ETTJAP.

Similar time lag results between the processes based on the RDUSAg, ETTUSA, and FFUSA are not observed here. However, an ADL approach together with the SRT showed a time lag of 4 years between the processes related to RDUSAg and FFUSA (results not shown). Again, this is consisted with the lag of 5 years observed here between the processes related to RDUSAg and ETTUSA.

5.3 Results from the autoregressive distributed lag (ADL) approach

In this section, the impact of R&D expenditures and GDP on Euro-Total filings is analyzed with the ADL approach.

For stationarity reasons, see Table 6.2, the input processes x and y and for the output process z were used, where

$$x = \nabla^1 \text{GDPg}, \quad y = \nabla^1 \text{RDTOTg}, \quad z = \nabla^2 \text{ETT}.$$

To identify a proper model, the ACF and the CCF belonging to the three processes are plotted in Fig. 6.4. In particular, these plots show from left to right row-wise in S-PLUS notation the ACF or CCF of these series:

Row 1: x and x, x and y, x and z
Row 2: y and x, y and y, y and z
Row 3: z and x, z and y, z and z

Studying these plots, one finds:

a. y is lagged by 0 and by 1 year to x, (second row, first plot).
b. z is lagged by 0 years to x, (third row, first plot).
c. z is lagged by 4 and 5 years to y, (third row, second plot).
d. z shows an autoregressive structure of order 1, (third row, third plot).

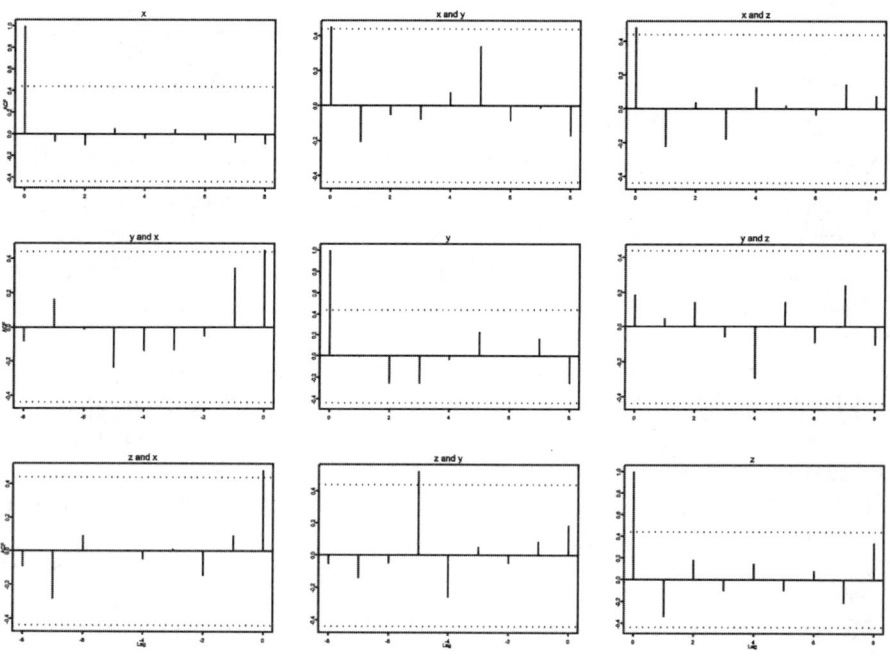

Fig. 6.4. ACF and CCF plots of the processes x, y, and z

Table 6.4. Estimated parameters of model M_1 with p-values of the corresponding t-test

Regressor	Parameter	p-value
Intercept	703.5555	0.3787
$x.lag0$	171073.5634	0.0951
$y.lag4$	-47079.2391	0.0834
$y.lag5$	41752.3381	0.1915
$z.lag1$	-0.2475	0.2712

Based on these dependencies, two models can be stated. In the first model, the (b)-(d) relations are taken into account, while relation (a) is not directly related to the output process z and therefore omitted. Overall, this results in:

$$(M_1): \quad z_t = \alpha_0 + \alpha_1 z_{t-1} + \alpha_2 x_t + \alpha_3 y_{t-4} + \alpha_4 y_{t-5} + \varepsilon_t,$$

where $(\varepsilon_t)_{t \in \mathbb{Z}}$ is a GWN process. In the second model, relation (a) is substituted into relation (c) to model the influence of y on z by x on z. This results in a long-memory GDP model. In particular, one gets

$$(M_2): \quad z_t = \alpha_0 + \alpha_1 z_{t-1} + \alpha_2 x_t + \alpha_3 x_{t-1} + \alpha_4 x_{t-4} + \alpha_5 x_{t-5} + \alpha_6 x_{t-6} + \varepsilon_t,$$

where, again, $(\varepsilon_t)_{t \in \mathbb{Z}}$ denotes a GWN process.

Fitting the first model to the data, perfect results for the error process (ε_t) w.r.t. GWN are obtained, i.e. the p-values of the Ljung-Box test for no autocorrelation and the normality-test are both above 0.9.

To simplify the model, the stepwise reduction technique, SRT, was applied. For this, we consider the p-values of the t-tests for every parameter (with null hypothesis: "parameter = 0") and calculate the AIC for every model, where the parameters are skipped successively according to their significance level. I.E. skip α_0 the intercept first, fit the new model, and skip the parameter with the highest p-value next, etc. Finally, the forecast is based on the model with the lowest AIC value, which is derived here instantaneously after the skip of the parameter belonging to y_{t-5}, i.e. after skipping the α_0 and the α_4 parameter in M_1. Thus, we get

$$(M_{1,a}): \quad z_t = \alpha_1 z_{t-1} + \alpha_2 x_t + \alpha_3 y_{t-4} + \varepsilon_t,$$

with estimated parameters and corresponding p-values given in Table 6.5.

This final model also passes the normality and the autocorrelation tests (p-value of Ljung-Box test: 0.82, p-value of normality test: 0.73).

Table 6.5. Estimated parameters of model $M_{1,\alpha}$ with p-values of the corresponding t-test

Regressor	Parameter	Value	p-value
z.lag1	α_1	-0.2761	0.2225
x.lag0	α_2	248361.7635	0.0082
y.lag4	α_3	-52109.9116	0.0577

It appears to be confusing that the parameter α_3, which belongs to y_{t-4}, is negative, since this means that an increase of y_{t-4} causes a decrease of z_t. However, it is not correct to conclude from this result, that an increase of the R&D growth rates causes a decrease of filings in general! The input series y is the one-time differenced R&D growth rate and therefore the result has to be read as

$$\alpha_3 \cdot y_t = -\alpha_3 \cdot (\text{RDTOTg}_{t-5} - \text{RDTOTg}_{t-4}).$$

Furthermore, the output series z is the two-time differenced ETT series and it is the impact of y on z that is considered here. To find a presentation of the results of the estimated model which is more suitably for interpretation, one has to transform x, y, z back to GDPg ($\equiv W$), RDTOTg ($\equiv R$), and ETT ($\equiv E$), respectively, and rearrange the terms properly to get

$$E_t \approx (2+\alpha_1) \cdot E_{t-1} - (1+2 \cdot \alpha_1) \cdot E_{t-2} + \alpha_1 \cdot E_{t-3} + \alpha_2 \cdot (W_t - W_{t-1}) + \alpha_3 \cdot (R_{t-4} - R_{t-5}).$$

Substituting the estimated values of the parameters α_1, α_2, and α_3 into this equation then shows how the developments of ETT up to time $t-1$, GDPg, and RDTOTg influence the filings of the year t according to the model $M_{1,\alpha}$. Note that an increase from R_{t-5} to R_{t-4} leads to a lower E_t. But note also that E_{t-2} appears in this equation with a negative sign, which says that a lower value of E_{t-2} leads to a higher value of E_t. From these two results we can conclude that the increase from R_{t-5} to R_{t-4} leads to an increase of E_{t+2}, which is reasonable.

To use the model $M_{1,\alpha}$ for prediction, one has to forecast the period 2001–2004 of GDPg first. In particular, we get the scenario given in Table 6.6, where the lower and upper prediction band is given by subtracting or adding two times the corresponding standard error, respectively.

The fit of M_2 also shows perfect results w.r.t. the GWN structure of the error terms. Using again SRT one finally gets

$$(M_{2,a}): \quad z_t = \alpha_1 z_{t-1} + \alpha_2 x_t + \alpha_3 x_{t-1} + \varepsilon_t,$$

as the model with the lowest AIC which also passes the normality and the autocorrelation tests (p-value of Ljung-Box test: 0.93, p-value of normality test: 0.71).

Table 6.6. Conditional forecast under $M_{1,\alpha}$, based on estimated GDPg and observed R&D growth rates

YEAR	GDPg	FORECAST	LOWER	UPPER	STD. ERR
2001	1.5%	156866	150040	163692	3413
2002	1.8%	173362	157982	188742	7690
2003	1.8%	187098	161206	212989	12946
2004	1.8%	201620	163555	239686	19033

Forecasting with this model, under the same estimated development of GDP growth rates for the period 2001–2004 as before, yields the results stated in Table 6.7.

Table 6.7. Conditional forecast under $M_{2,\alpha}$, based on estimated GDPg and observed R&D growth rates

YEAR	GDPg	FORECAST	LOWER	UPPER	STD.ERR
2001	1.5%	157327	150243	164411	3542
2002	1.8%	170861	154655	187067	8103
2003	1.8%	184155	156517	211793	13819
2004	1.8%	197565	156532	238599	20517

Both models, $M_{1,\alpha}$ and $M_{2,\alpha}$, are special cases of

$$z_t = \alpha_1 z_{t-1} + \alpha_2 x_t + \alpha_3 x_{t-1} + \alpha_4 y_{t-4} + \alpha_5 y_{t-5} + \varepsilon_t.$$

But analyzing this model with the above indicated stepwise parameter reduction technique results again in $M_{1,\alpha}$. Thus, no further improvement can be established with this extension.

Analyzing these TS with lower degrees of differencing shows autocorrelation structures which can hardly be interpreted reasonably. This is mainly caused by the fact that the series are then not stationary anymore and therefore the estimated autocorrelation is not guaranteed to be consistent.

Instead of the growth rates of the GDP index and R&D expenditures, now these series are used directly, but still properly differenced. In particular, the processes

$$x = \nabla^2 \text{ GDP}, \quad y = \nabla^2 \text{ RDTOT}, \quad z = \nabla^2 \text{ ETT},$$

are considered. Based on the ACF and CCF plots (not given here) the model

$$(M_3): z_t = \alpha_0 + \alpha_1 z_{t-1} + \alpha_2 x_t + \alpha_3 x_{t-1} + \alpha_4 y_{t-4} + \alpha_5 y_{t-5} + \varepsilon_t,$$

seems to be appropriate, where $(\varepsilon_t)_{t \in \mathbb{Z}}$ denotes a GWN process. Applying again the stepwise parameter reduction approach, one finally ends up with

$$(M_{3,a}): z_t = \alpha_1 z_{t-1} + \alpha_2 x_t + \alpha_3 y_{t-4} + \varepsilon_t$$

as the best choice.

The forecasts based on this model are given in Table 6.8.

A summary of the conditional forecasts obtained from these three models, together with the actual development of Euro-Total filings, is given in Table 6.9.

Comparing these three forecasts with the actual numbers of filings, model $M_{3,\alpha}$ produces the best predictions w.r.t. the standard errors and also w.r.t. the deviation to the actual filings for the period 2002–2004. However, the forecast for the year 2001 is too conservative, since the actual filings of this year are above the upper prediction band. Because 2001 is the first year to be forecasted, this should rule out $M_{3,\alpha}$ as a good candidate, generally. Furthermore, taking into account the unpredictable collapse of the new economy in 2001 and the unexpected terrorist attack of 11 Sep. 2001, which both had the effect of forcing the development of Euro-Total filings to decline, the bad forecast of Euro-Total filings in 2001 is even worse and therefore model $M_{3,\alpha}$ is unacceptable.

The $M_{1,\alpha}$ and $M_{2,\alpha}$ forecasts show all the actual filings within their prediction bands, where the $M_{2,\alpha}$ based forecast fits better than the $M_{1,\alpha}$ based forecasts to the actual development, while the $M_{1,\alpha}$ forecasts have smaller standard errors.

Table 6.8. Conditional forecast under $M_{3,\alpha}$ based on estimated GDP and observed R&D

YEAR	GDPg	FORECAST	LOWER	UPPER	STD. ERR
2001	1.5%	154360	147957	160763	3201
2002	1.8%	169351	154944	183757	7203
2003	1.8%	181052	156825	205279	12113
2004	1.8%	193528	157937	229118	17795

Table 6.9. Standard error and deviation of ETT actual and forecast per year in % to forecast obtained under $M_{1,\alpha}$, $M_{2,\alpha}$, and $M_{3,\alpha}$

		$M_{1,\alpha}$		$M_{2,\alpha}$		$M_{3,\alpha}$	
YEAR	EURO-TOTAL	DEV.	STD.ERR	DEV.	STD.ERR	DEV.	STD.ERR
2001	162020	3.29%	3413	2.98%	3542	4.96%	3201
2002	161068	-7.09%	7690	-5.73%	8103	-4.89%	7203
2003	167205	-10.63%	12946	-9.20%	13819	-7.65%	12113
2004	178843	-11.30%	19033	-9.48%	20517	-7.59%	17795

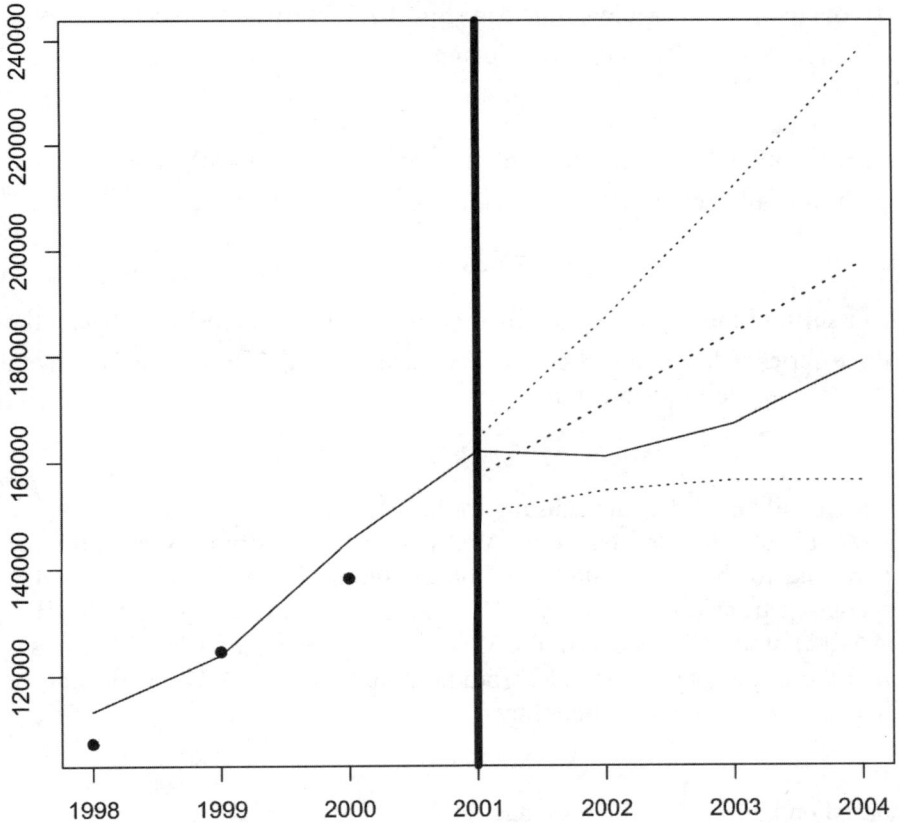

Fig. 6.5. Fit, forecast, and prediction bands of ETT based on model $M_{2,\alpha}$. Actual figures are counted from the database in 2005

5.4 Results from the vector autoregressive (VAR) approach

In this analysis we fit a VAR model to the growth rates of R&D expenditures, to the growth rates of GDP, and to Euro-Total filings. For stationarity reasons, we have to consider the processes x, $\bar{y}$, and z defined as

$$x = \nabla^1 \text{GDPg}, \quad \bar{y} = \nabla^1 \text{RDTOTg}, \quad z = \nabla^2 \text{ETT}.$$

Since there is a delay of 4 or 5 years until $\bar{y}$ affects z, it is necessary to fit a VAR model here with at least 5 lagged terms in order to get reasonable results. This, however, is not possible due to the small number of observations of the z-series compared to the large number of model parameters. To overcome this problem, one can shift the $\bar{y}$-series by 4 years and use a VAR(1) or VAR(2) model. Now denote by

$$y_t = \bar{y}_{t-4},$$

then the model is build upon the processes (x, y, z) with "observations" per time unit t given by

$$\left(x_t, \ y_t \equiv \bar{y}_{t-4}, \ z_t \right).$$

But shifting the y-series, the natural year-based relationship between the three series is lost and can not be identified in the data anymore. However, it is reasonable to assume that

$$y_t \to x_t \to z_t$$

which will be used as the causality order later on.

As already pointed out, only VAR(1) or VAR(2) models are possible here due to the small number of observations. From these two possible models it turns out that the VAR(1) model fits better to the data than the VAR(2) model. Precisely, in the VAR(2) model the residuals of the z-series show significant departure from normality. Thus VAR(1) is the choice here and we denote this model by

$$(V): \text{VAR}(1)$$

based on x, y, and z without intercept.

Table 6.10 shows the estimated parameters together with the corresponding p-values of the t-test. Note that y_lag1 equals $\bar{y}_{t-5}$. Significant departures from the null hypothesis (parameter = 0) can be observed in this table. In particular, we find that $\bar{y}_{t-5}$ has an impact on x_t and, more important for our study, an impact on z_t. Furthermore, z_t is also influenced significantly by z_{t-1}. Fortunately, no significant impact of x_{t-1} or z_{t-1} on $\bar{y}_{t-4}$ can be found in the table, since this would indicate an influence of the current GDP or ETT figures on RDTOT five years ago.

Table 6.10. Estimated parameters of V with p-values of corresponding t-test

	x- process		y- process		z- process	
Regressor	Parameter	p-value	Parameter	p-value	Parameter	p-value
$x.lag1$	-8.832×10^{-2}	0.71	-1.002×10^{-0}	0.25	100586.2	0.31
$y.lag1$	1.744×10^{-1}	0.03	3.044×10^{-2}	0.91	64062.6	0.05
$z.lag1$	-2.600×10^{-7}	0.70	1.280×10^{-6}	0.60	-0.4	0.12

The residuals of the fitted V model passed, separately for each process, the autocorrelation and the normality test. Furthermore, the stability test is also passed by this model.

The propagation over time of a single, orthogonal shock (innovation) within this model is shown in Fig. 6.6.

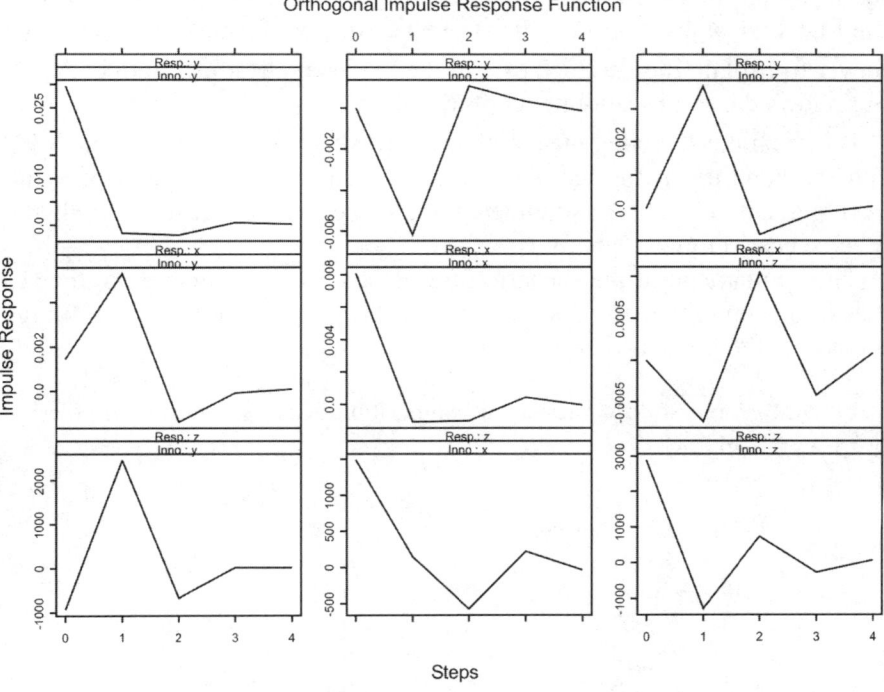

Fig. 6.6. Orthogonal impulse-response function of V w.r.t. the causal ordering $y \rightarrow x \rightarrow z$

The last row of plots shows the impact of these shocks on the process z. An innovation due to the y-series starts with an immediate (negative) effect on the z process. Then, after one time unit we observe a large positive effect which is absorbed after two time steps. Again we find an indication of the impact of $\bar{y}_{t-4}$ and $\bar{y}_{t-5}$ on z_t. Furthermore, the impulse-response analysis shows an immediate response of z due to an innovation of x which flattens out after one time unit. Finally, the obvious immediate reaction of z to an innovation of z can be observed in the right plot of the last row. This impulse also shows a reaction after one time step and turns down in the following steps.

The decomposition of the forecast error variance is shown in Fig. 6.7 and listed under Table 6.11. The variance of the forecast error for the one year ahead prediction is mainly caused by the variance of the errors itself (73%) and of x (20%), while the variance of the error y has no significant impact here (7%). For the two year ahead prediction, the situation changes. Now the innovation of y accounts for 36% of the total variances of z, while the impact of the innovations of z drops down to 52%. Combining these results for the first two years of prediction, it can be concluded that the first year innovation of y has to be responsible for this increase in the power to explain the variance of z in the two step ahead prediction. Again this shows the significant impact of $\bar{y}_{t-5}$ on z_t.

Based on the V model, the simultaneous forecast is shown in Table 6.12, and the conditional (or scenario) forecast in Table 6.13, where the same development of GDP_g is assumed for the scenario forecast as in all the other scenario forecasts before.

The comparison of the scenario forecast and the development of the actual filings up to the year 2004 is given in Table 6.14. All the actual filings are within the prediction bands.

Table 6.11. Forecast error variance decomposition of z w.r.t. the causal ordering $y \to x \to z$ of model V

Forecast horizon	Proportion due to		
	y	x	z
1	0.0764	0.1935	0.7301
2	0.3610	0.1167	0.5222
3	0.3599	0.1252	0.5149
4	0.3578	0.1268	0.5154
5	0.3577	0.1269	0.5155

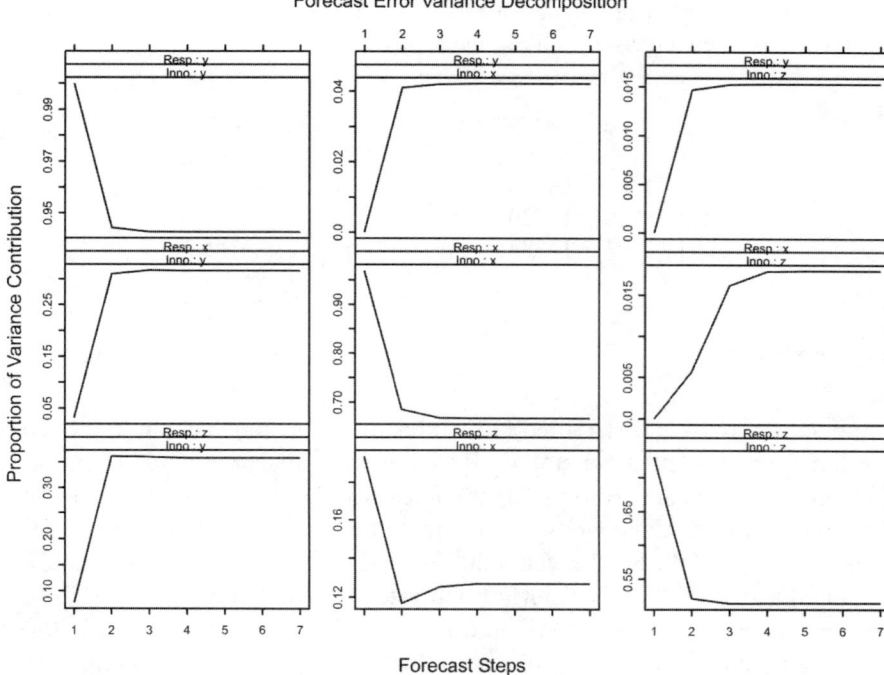

Fig. 6.7. Forecast error variance decomposition of V w.r.t. the causal ordering $y \to x \to z$

Table 6.12. Simultaneous long-term forecast based on V

YEAR	ETT	GDPg
2001	161887	0.037
2002	181541	0.037
2003	197369	0.033
2004	216294	0.039

Table 6.13. Conditional VAR-Forecast under the V model based on estimated GDP and observed R&D

YEAR	GDPg	FORECAST	LOWER	UPPER	STD. ERR
2001	1.5%	157730	151521	163938	3104
2002	1.8%	173315	159276	187354	7019
2003	1.8%	186301	162594	210009	11854
2004	1.8%	201145	166189	236101	17478

Table 6.14. Standard error and deviation of ETT actual and forecast per year in % to forecast obtained under V (with estimated GDPg)

YEAR	EURO-TOTAL	V	
		DEV.	STD. ERR
2001	162020	2.72%	3104
2002	161068	-7.07%	7019
2003	167205	-10.25%	11854
2004	178843	-11.09%	17478

6 Conclusions

Some autoregressive distributed lag models under a stepwise parameter reduction approach and a general VAR model were analyzed here. The models were evaluated according to their forecasting ability and their precision. To compare the models, we had to estimate the development of the growth rates of global GDP, for the years 2001–2004. Since this part of the analysis was done in 2003, the exact growth rates for the years 2001 and 2002 (as they were known in 2003) could be used. The growth rates for 2003–2004 are assumed to be the same as in 2002. Based on this evaluation, we found 3 models that were adequate for forecasting purposes. In Table 6.15 the structures of these models are summarized, where we use

$$x = \nabla^1 \text{GDPg}, \quad y = \nabla^1 \text{RDTOTg}, \quad z = \nabla^2 \text{ETT}$$

Table 6.15. Dependence structure of the analyzed forecast models

Model	Forecast of z_t is based on				
	x_t	x_{t-1}	y_{t-4}	y_{t-5}	z_{t-1}
$M_{1,\alpha}$	×		×		×
$M_{2,\alpha}$	×	×			×
V		×		×	×

The deviations of the actual development of Euro-Total filings up to 2004 to the different forecasts and the standard prediction errors of the forecasts are listed in Table 6.16.

All the three models overestimate the actual filings for the years 2002–2004. But there have been several unexpected and unpredictable occurrences during the period 2001–2004 that forced the number of filings to decrease. The actual filings are within the prediction bands of all three models, and these bands are narrowest for the VAR model.

The number of filings in 2001 is better predicted by the V model than by the M ones. Furthermore, for the one year ahead prediction, one gets even better results for the VAR model if no assumption on the further development of GDPg at all is made, as Table 6.17 shows. Nevertheless, the unconditional forecasts due to the VAR model are poor for 2 and more years ahead prediction. It is remarkable that the one year ahead forecast based on the VAR model is substantially better than those which are based on the M models, since the M models also incorporate the GDPg of the year which has to be forecasted, while the VAR model takes only the information of GDPg up to one year before, see Table 6.15.

The models $M_{1,\alpha}$ and V use information of GDPg and RDTOTg, while $M_{2,\alpha}$ is based on GDPg alone. Thus the latter model is more easy to use w.r.t. data collection, since RDTOTg is a more complicated data set than GDPg and it might be not available by the time the forecasts have to be produced. On the other hand, all these conditional forecasts are based on preestimated GDPg for the period where the filings have to be predicted. This preestimated GDPg might be wrong. In this case, however, $M_{1,\alpha}$ and V should be better than $M_{2,\alpha}$, since the first two are also based on the known development of RDTOTg.

Table 6.16. Standard prediction error and deviation of ETT actual and forecast per year in % to forecast obtained under $M_{1,\alpha}, M_{2,\alpha}$, and conditional V (with preestimated development of GDPg)

YEAR	EURO-TOTAL	$M_{1,\alpha}$		$M_{2,\alpha}$		V	
		DEV.	STD.ERR	DEV.	STD.ERR	DEV.	STD.ERR
2001	162020	3.29%	3413	2.98%	3542	2.72%	3104
2002	161068	-7.09%	7690	-5.73%	8103	-7.07%	7019
2003	167205	-10.63%	12946	-9.20%	13819	-10.25%	11854
2004	178843	-11.30%	19033	-9.48%	20517	-11.09%	17478

Table 6.17. Forecast, standard prediction error, and deviation of ETT actual in 2001 under (unconditional) VAR model V without further assumptions on the development of GDPg

	Model V			
YEAR	EURO-TOTAL	FORECAST	DEV.	STD.ERR
2001	162020	161887	0.08%	3547

7 Annex

In the following list, most of the symbols that were used in the article are explained.

a = b	a is equal to b.
a ≤ b	a is less or equal to b.
a ≥ b	a is greater or equal to b.
a < b	a is less than b.
a > b	a is greater than b.
a<<b	a is much less than b.
a>>b	a is much greater than b.
a ≈ b	a is approximately the same as b.
a ≡ b	a and b are equivalent notations.
a ∈ A, or A ∋ a	a is an element of the set A.
$f : A \ni a \to f(a) =$ "definition"	The elements a of the set A are mapped by the function f to $f(a)$, where the meaning of $f(a)$ is given by "definition".
$\mathbf{Z}$	Set of integers.
$\mathbf{Z} \times \mathbf{Z}$	Integer grid in the plane.
$\mathbf{Z}^d$	d-dimensional space of integers.
R	Real numbers.
R^d	d-dimensional Euclidean space.
~	Distributed as.
$X \sim N(\mu, \sigma^2)$	The random variable X is distributed as a normal random variable with expectation μ and variance σ^2.
$\text{GWN}(0, \sigma^2)$	Centered Gaussian white noise process with variance σ^2.
$\text{WN}(0, \sigma^2)$	Centered white noise process with variance σ^2.
$E(X)$	Expectation of the random variable X.
B	Backshift operator.
∇	Difference operator.
$\sum$	Summation sign.

Chapter 7

International patenting at the European Patent Office: aggregate, sectoral and family filings

Walter G Park

Department of Economics, American University, Washington, DC, US

Modules: C, D
Software used: STATA

1 Introduction

This chapter provides a panel data perspective of patent filing behavior at the European Patent Office (EPO). The EPO filings of different source countries are observed and analysed over the period 1980–2000. Moreover, forecasting exercises are conducted for different aspects of patent filings at the EPO. In particular, the chapter examines the behavior of total EPO patents as well as patents disaggregated by mode of filing, technological sector, and selected patent families. Analysis of patenting behavior reveals the nature of the underlying demand for patents. Research and development (R&D) is an important influence on both the propensity to file patents and the potential pool of inventive output. In terms of forecasting performance, the analysis finds that a dynamic model augmented with R&D generally performs best (based on root mean squared proportion errors as measures of forecast accuracy). The study includes examples of some sample forecasts for individual source countries.

Nations trade and invest physical capital in each other's markets. Indeed, the international economy has become much more interdependent through these trade and investment linkages. Less well understood, however, is the increased interdependence due to the diffusion of technological ideas among nations. The international patent system and institutions governing intellectual property rights help support a formal marketplace for knowledge capital. Yet the volume, direction, and underlying determinants of international patent flows have not been the subject of much inquiry.

Thus far, few studies exist that seek to explain and forecast patent filings. In general, these studies treat the nation as the unit of analysis, and focus on whether the recent growth of patenting is primarily innovation-driven or due to the strengthening of patent laws. What has not been addressed is the global breadth of patenting activities. Moreover, due to their national perspectives, the existing literature pays scant attention to regional or multilateral patenting systems (among a bloc of nations), such as that of the EPO. Such systems are relevant to accounting for the world wide growth and spread of patenting.

Two factors motivate this study. First, the increasing prominence of regional and supranational offices, such as the EPO or WIPO, has fundamentally changed the way inventors obtain patent protection. This study focuses on analyzing the increased worldwide demand for EPO patents and uses the conceptual models of patenting behavior to assess their ability to explain and predict EPO patent filings. An improved understanding of EPO filing behavior is an important step towards characterizing the growth in world wide patenting.

A second motivation is that, for national and regional patent offices alike, the extent of patenting activity has implications for internal workload (processing applications, conducting searches and examinations, and so forth) and patent office revenues (which are determined, among other things, by the volume of filings and official fees). A better understanding of the underlying determinants of the demand for patents could better assist organizations like the EPO to price its services, project revenues, and make operational decisions. Improved projection of patenting demand could be useful in any work-sharing or revenue-sharing arrangements with national offices or with other supranational offices. For instance, WIPO administers the Patent Cooperation Treaty (PCT) which provides, for among other things, a system of international patent applications (WIPO 2006a). Thus trends in euro-direct filings versus euro-PCT-IP filings would be useful for coordination and workload planning between the EPO and WIPO.

This chapter is organized as follows: the next section provides a brief literature review. Section 3 discusses the empirical framework and methodology. Section 4 presents some forecasting exercises conducted with the basic patenting model. Three kinds of patent filings will be the subject of forecasts: firstly, aggregate patent filings at the EPO. The term *aggregate* here refers to the sum of filings across technological fields. Furthermore, the breakdown of these filings by mode of filing is considered; that is, whether the filing is euro-direct or euro-PCT-IP, and whether the filing is a first filing or a subsequent filing.

The second type of filings for which forecasting exercises will be conducted are sectoral filings. The term sector will refer to the field of technology (classified according to the EPO's joint cluster divisions). The third (and last) type of filings that will be forecast are patent family filings. The definition of a patent family that will be used has been introduced in Chap. 1, "a group of patent filings that claim the priority of a single filing, including the original priority forming filing itself and any subsequent filings made throughout the world" (EPO, JPO, USPTO 2005). First filings are priority forming applications that do not claim the priority of any previous filing, while subsequent filings constitute all other applications. The latter are usually made within one year of the first filings, because of the stipulations of the Paris Convention (WIPO 2006). A distinct set of priority forming filings is used to index the set of patent families. From the data set on international patent family filings, those families that contain a subsequent filing at the EPO (and/or other types of 'Blocs') can be selected. Finally, a concluding section will summarize the main results and discuss some extensions for further study.

Overall, this study finds that EPO patenting is significantly driven by R&D activities, and that forecast accuracy is generally improved through the use of R&D along with dynamic terms representing lagged patenting. Forecast performance does vary somewhat by technological field, by mode of filing, and by nature of patent family. The good forecasts can, in some cases, come within 90–95% of actual filings. The forecast accuracies are not too sensitive to the methods of estimation considered.

2 Literature review

Relative to the literature at large on the economics of the patent system, very little empirical work to date exists on the determinants of patenting, and none with a specific focus on regional patenting systems, such as that of the EPO. There are studies on the impacts of the patent system (on innovation, trade, productivity, and welfare), but not very much on what drives patenting behavior.

First, one set of studies is based on firm level surveys (interviewing managers as to why firms patent and as to how important patents and patent laws are to the firms); the second set is based on statistical data sources (conducting regression analyses on patent data in order to infer the factors that influence patenting).

As a prelude to the survey studies, the conventional wisdom had been that firms demand patent protection in order to safeguard their intangible assets, which are easy to copy and distribute at nearly zero marginal cost (without other producers needing to incur any of the 'sunk' development costs). Infringement and imitation work to dissipate the gains to firms and thereby (ex ante) reduce their incentives to innovate. Recent surveys have challenged head on whether patent protection is necessary to stimulate investment in invention and commercialization. The Levin et. al. (1987) survey of US firms' patenting behavior reports findings which have generated much controversy – namely that firms do not, in general, regard patent protection as very important to protecting their competitive advantage (and thus to appropriating the returns to their investments). The idea is that firms have various alternative means (other than patenting) for appropriating the rewards to their innovations; for example, trade secrecy, lead time, reputation, sales and service effort, and moving quickly down the learning curve. Patent protection ranked low among these alternative means of appropriation. The study therefore questions previous understanding of what motivates patenting.

The question then is, if a patent is not important as an instrument for appropriating the returns to innovation, why do firms patent (and patent a lot)? The survey by Cohen et. al. (1997) reports that firms have various reasons to patent – as a means to block rivals from patenting related inventions, as strategic bargaining chips (in cross-licensing agreements), as a means to measure internal performance (of the firms' scientists and engineers), and so forth. Thus these various other factors are what primarily determines (or motivates) patenting, rather than the protection of their R&D investment returns.

Some criticisms can be made of these survey analyses. Firstly, it would be useful to update the sectors under study to incorporate new industries which have emerged since the surveys were conducted. The biotechnology and software industries may, for example, provide interesting perspectives on the rationale for and importance of patenting. Secondly, the responses of firms (or their attitudes towards patents) may have been influenced by the patent regime in place. It would be useful to separate these two out. Thirdly, the responses of interviewees may not be fully comparable. One person's rating of 9 out of 10 may differ from another's. There is no anchor in the way that ratings are scaled. Thus it is difficult to tell whether the responses reflect differences in firm behavior or random errors. Finally, while the surveys are very time-consuming and commendable work, the information is based on US firms' experiences. A similar comprehensive study for Europe and Asia, and so forth, would shed more light on patenting behavior – such as why firms patent globally and if so, why they choose certain routes (e.g. EPO, PCT, etc.).

Among the statistical database studies, Schiffel and Kitti (1978) is one of the earliest works. This study was motivated by the fact that foreign patenting in the US, during the period 1963–73, grew at a faster pace than US patenting abroad. This seemed to have created concerns about the loss of US technological leadership, a conclusion which the authors challenged. The study finds that the rise in foreign filings in the US reflected increased world trading opportunities, and not a reduction in US inventiveness vis-à-vis foreigners.

Bosworth (1980) examines a larger sample of countries using cross-sectional data. Bosworth finds that certain patent law features do not explain US patenting abroad. This is at odds with the strong advocacy US firms have shown towards international intellectual property law reform. Later work (as described below), which improves upon the measurement of patent regimes, does show the importance of patent rights to international patenting behavior (including US patenting abroad). Bosworth (1984) repeats the analysis for patenting flows into and out of the UK, and finds qualitatively similar results.

Slama (1981) fits a gravity model to international patenting data (for 27 countries during the pre-EPO period, 1967–78). The dependent variable is cross-country patenting as a function of the GNPs and populations of the country of origin and destination, the geographic distance between (capital cities of) countries, and dummy variables for regional trade membership. A key finding is that regional trade areas create positive preference, in that members engage in more bilateral patenting than would otherwise be the case.

In contrast to the previous studies, Eaton and Kortum (1996) develop a decision-theoretic model of patenting. They use this model patenting behavior to explain some of the sources of differences in productivity across countries, namely to impediments in the diffusion of technology (measured via flows of international patent filings), which would otherwise enable countries to catch up technologically.

Park (2001) studies the extent to which international technology gaps, as measured by total factor productivities, can be explained by differences in patent protection levels and patenting across countries. The focus is on whether international patent reform helps narrow technology gaps. The study finds that patent reforms alone have modest impacts on narrowing technology gaps in the short run due to the fact that, in the short time horizon, patent reform largely stimulates the filings of patents of marginal value.

In other studies, the focus of attention is the trend in patenting itself. Kortum and Lerner (1999) observe an 'explosion' in US patenting (domestically and abroad) and examine several hypotheses that might explain that. The two critical competing hypotheses are the pro-patent hypothesis and the fertile technology hypothesis. According to the pro-patent (or friendly court) hypothesis, changes in the legal regime precipitated the increase in patenting (for example, via the establishment of a specialized appellate court called the Court of Appeals of the Federal Circuit, which appeared to render decisions favorable to patent holders, upholding patent validity decisions or reversing invalidity rulings). This increased the incentive to acquire patent rights. According to the fertile technology (or increased inventiveness) hypothesis, firms have become more productive and the management of R&D more efficient – hence the rise in patent applications. In a sense, the two hypotheses are not altogether separable. To the extent that strengthened patent rights stimulate R&D, the regime changes might have led as well to increased innovation potential. Secondly, increased R&D efficiency and innovation potential might have been the reason the courts ruled more favorably to patent rights holders; patents awarded to higher quality technologies would less likely be ruled as invalid. Thus, it is not clear that two distinct hypotheses are being examined.

A micro-level study by Hall and Ziedonis (2001) challenges the hypothesis that US firms patented more because they were more inventive. Using a sample of US semiconductor firms, the authors find that the motive for, or determinant of, patenting is strategic: to pre-empt "hold-ups" or blocking if rivals own key patents. The argument is that if a firm could own critical patents itself, it could better negotiate with others who have rights to technologies that the firm might need. The authors argue that recent legal changes put firms in a situation where they need to patent for this purpose. The legal changes broadened patent scope and facilitated entry by specialized firms. In an environment of cumulative innovation (such as in the semiconductor industry), the possibilities for patent hold-up are greater. Firms can not afford not to acquire patents while others are amassing vast patent portfolios. The filing of these vast patent portfolios may account for the explosion in patenting in recent years.

To summarize, the existing literature suggests a variety of motivations for patenting for addressing particular policy issues (such as the merits of patent reform). The research agenda has been focused on explaining and testing specific hypotheses about patenting behavior rather than on developing models that have predictive value; that is, models that can provide good forecasts of trends in patenting behavior. Ultimately, a useful test of models of patenting behavior is how well they predict real world patenting behavior. In general, models without a dynamic specification (or that do not yield lagged adjustments in patenting) fail to forecast well, which would cast doubt on whether the models fully capture the underlying processes driving patenting behavior.

3 Methodology and data sets

As emphasized in the introduction, good forecasts of patent filings (by technology and/or by mode of filing) are useful to the EPO for purposes of allocating internal resources. For the EPO's external relations with other patent offices, coordination of tasks is enhanced by good forecasts of the breadth of international patent filing activities (whether they involve two or more countries, or blocs of countries). Three types of forecasting exercises are conducted:

- Overall patent filings in the EPO, broken down by modes of filing; in particular, applicants can file patents directly at the EPO or indirectly via the PCT. Furthermore, these filings may be 'first filings' or 'subsequent filings'.
- Patent filings broken down by technological field. The technological classification adopted here is that of the EPO joint cluster (JC) system, which consists of fourteen technological units (e.g. unit 1 is electricity and electrical machines, unit 2 is handling and processing, etc.). The EPO examining divisions consist of directorates assigned to particular JC's.
- Patent filings comprising patent families. The EPO patent family database PRI is indexed by priority forming filings and provides related subsequent filing activity in the major blocs: EPC (including the EPO), US, Japan, and Others. The database thus enables the user to pick out the type of patent families one seeks to examine.[1]

Each of these exercises will be considered in turn. For each forecasting exercise, there will be a discussion of some recent trends in filings, regression estimates, forecast accuracy, and sample forecasts for a given year (by individual source countries).

3.1 Conceptual framework and methodology

3.1.1 Static view

This section builds on the conceptual framework developed in Chap. 3, Sect. 2. Consider the following model of patenting behavior:[2]

$$P_{ij} = \alpha_i \, \delta_{ij} \, f_{ij} \qquad (1)$$

---

[1] For more details on the EPO patent family statistics database PRI, see Hingley and Park (2003). This study draws upon material in that earlier paper.
[2] See Chap. 3 of this volume for the microfoundations underlying this model.

where P_{ij} denotes patent applications from source country i in destination country j, α_i the flow of patentable innovations (in source country i), δ_{ij} the fraction of α_i that has applicability in destination j, and f_{ij} the fraction of $\alpha_i \delta_{ij}$ that is applied for patents in destination j. The pool of patentable innovations (in a given period) should depend on the extent of research and development (R&D) activity, while the propensity to patent them in a given destination should depend on the attractiveness of the destination market. The cross-country applicability of innovations should depend on bilateral factors, which will be treated as country-pair specific random effects. The propensity to patent, f_{ij}, should depend on whether the value of patenting exceeds the cost. The value of patenting should be the difference between the rewards to an inventor from patenting an innovation and the rewards from not patenting that innovation (say the default reward). In other words, patent applicants should be motivated by the increment in reward from patenting (relative to the cost).

3.1.2 Forecasting

Suppose patent applications are a function of some independent variable x:

$$P_{it} = \beta_0 + \beta_1 x_{it} + \varepsilon_{it} \quad (2)$$

where t = 1, ..., T denotes time (sample period), i = 1, ..., N denotes source countries, j = EPO (hence subscript j is omitted), and where $\varepsilon_{it} = \upsilon_i + \mu_{it}$ is the error term. In the panel dataset below, υ_i is used to capture the bilateral specific effect between a source country and the EPO destination.

Given (future values)

$$x_{iT+1}, x_{iT+2}, \ldots, x_{iT+k},$$

the estimates $\hat{\beta}_0, \hat{\beta}_1$ can be used to generate predictions for each source country i = 1, ..., N.

$$\hat{P}_{iT+1}, \hat{P}_{iT+2}, \ldots, \hat{P}_{iT+k}$$

Note that in the actual estimation and forecasting below, the dependent variable will be the natural logarithm of patent applications per (source country) worker. Thus, the above methodology needs to be modified slightly to take the exponent of P and multiply by number of workers to obtain the predicted number of patent applications (in natural units).

3.1.4 Forecast accuracy

Among the different criteria that could be used, this paper evaluates forecast accuracy by examining the Root Mean Square Proportion Errors (RMSPE).

For date T+k, given actual patent applications P_{iT+k} and predicted $\hat{P}_{iT+k}$:

$$\text{RMSPE}_{iT+k} = \sqrt{\left(\frac{P_{iT+k} - \hat{P}_{iT+k}}{P_{iT+k}}\right)^2}$$

The empirical section below provides mean RMSPE across source countries $i = 1, \ldots, N$ for different k-step ahead periods, as well as provides some sample forecasts by individual source countries.

To get an anchor for the root mean square proportion error, note that RMSPE = 1 if the predicted value is either twice that of the actual value or equal to zero. In other words, it gives us an idea of the proportion deviation from actual.

3.1.5 Dynamics

Suppose patent applications depend on past applications. Then an extension to Eq. (2) is:

$$P_{it} = \beta_0 + \beta_{11} P_{it-1} + \ldots + \beta_{1j} P_{it-j} + \beta_2 x_{it} + \upsilon_i + \mu_{it}, \tag{3}$$

Through the lagged variables, the entire history of the dependent variables P_{it} is reflected in the equation. Thus the effect of the independent variable x is conditioned on this history. The impact of x on P reflects the effect of *new* information.

For comparison, the autoregressive models (AR1 and AR3) – i.e. models without x's – are examined. Typically the x's will be measured by the logarithm of real research and development (R&D) expenditures of the source country per source country worker. The outcomes of using other independent variables will also be described below. For comparison, the paper provides estimates of the above dynamic equation using generalized least squares (GLS) and generalized method of moments (GMM). The GLS here is random effects estimation and the GMM the Arellano and Bond (1991) method. The presence of lagged variables in the panel data introduces correlations between the right-hand side variables and the error term, for which differencing and instrumental variables are used to handle this problem. The Annex provides a brief review of these estimation methods.

The regression models tended to produce much better forecasts if lagged values of the dependent variable are included, given the serial correlation or momentum in patent filings over time. The reason for the primary focus on R&D as the independent variable of interest is that other variables such as output are correlated with R&D (since output is a function of R&D, among other factors). Hence R&D is both important in itself and acts as a proxy for other important factors.

The R&D variable, however, represents a source country characteristic. It should be noted, though, that the characteristics or attributes of the EPO tend largely to vary not across source countries but over time. In other words, the source countries all face (largely) the same conditions in the EPO destination (whether it be EPO policy, institutional factors, rules, market size, market conditions, and so forth). Thus most of the variation between the EPO filings of different source countries is likely to be due to source country factors. Nonetheless, developments in the EPO do occur over time that could stimulate or decrease the patenting of source countries (though not necessarily in the same way or to the same extent), and it would therefore be useful to develop proxy measures of EPO destination characteristics. However, in preliminary analyses, some difficulties were encountered in defining and deriving EPO destination variables (e.g. weighting and aggregating the member country characteristics). If the destination were a single country, this is easy to do. But for a bloc (such as the EPC contracting states that together run the EPO) one needs a measure of the market size or other characteristics of the bloc as a whole, and then to weight the underlying individual countries comprising the bloc. While the development of these variables is a work in progress, the dynamic lagged dependent variables may proxy for time shifts in conditions in the destination EPC contracting states.

Lastly, it would be useful to discuss the possible lag structure of R&D in relation to the effects on patenting. In preliminary analyses, the results were not qualitatively different if the first, second, or third lags of R&D flows are used. This may be due to a couple of factors. First, R&D itself is correlated with past values, reflecting the fact that the R&D behind an innovation is not a one-shot investment but part of a cumulative effort, which is why the stock of R&D was important. Secondly, while a given period's R&D may yield a patentable innovation with a lag, this is not to say that current R&D cannot influence current patenting activity. Firms may wish to file for patents before further refining their research projects or devoting more resources to them. The priority right gives added security and incentive to continue their R&D, if only to acquire proprietary rights to early versions of their innovations. Thus, R&D activity may stimulate current patent filing activity to the extent that firms take anticipatory action and seek priority rights to forthcoming innovations.

3.2 Data sources

A panel data set is used to estimate Eq. (3). The sample consists of 53 source countries over the period 1980–2000, and one destination, namely the EPO. Given a number of missing observations (due to incomplete data on the independent variables of interest), in practice this sample reduces to about 30 source countries over a 21 year period (providing 630 observations = 21 x 30). Data on EPO patent applications are from the European Patent Office. Data on national research and development (R&D) are from the OECD's Main Science and Technology Indicators database.[3] The main measure of R&D used is Gross Expenditure on Research and Development (GERD). This is a broad country-wide measure, encompassing R&D funded by industry, government, and non-profit sectors (such as universities). Due to knowledge spillovers, innovative activity is likely to depend on a broad stock of knowledge, not limited to industrially funded and performed research. In the sectoral sample, however, the measure of R&D used is Business Enterprise Research and Development (BERD) expenditures (funded by various sources, including government and industry).

For non-OECD and/or developing countries, R&D data are taken from UNESCO's Statistical Yearbook (UNESCO 1980–2002). For a number of countries, data are missing. For data that were missing between years, we filled in gaps by a linear interpolation.[4]

Table 7.1 shows some sample means (over the period 1980–2000) of patent filings at the EPO and R&D as a percentage of GDP by country group. The countries in the sample are grouped according to their sample average real GDP per capita, using World Bank (2002) figures. The high-income group refers to those countries whose GDP per capita exceeded $ 17000 (real 1995 US dollars); the low-income group to those whose GDP per capita was less than $3500 (real 1995 US dollars); and the medium-income group to those whose GDP per capita was between those two limits. For each income group, the within-group mean values are provided, and at the end of the table, the overall mean values of the variables are

---

[3] See http://www.sourceoecd.org, June 2002.
[4] For example, if there were a three year gap in R&D flows, the total change in R&D values over that period would be divided by three and the value of R&D would then be incremented by that figure for each year that was missing. For example let $\Delta = (x(t+3) - x(t))/3$, where t denotes time. Then $x(t+1) = x(t) + \Delta$, $x(t+2) = x(t+1) + \Delta$.

provided. In general, the high-income nations do most of the patenting in the EPO. The high-income nations generally have the highest rates of R&D spending (averaging 1.53% of GDP), while the low-income nations have an average R&D to GDP ratio of 0.48%.

Table 7.1. Sample statistics of EPO patenting and R&D-to-GDP: Average 1980–2000

	EPO Patent Filings	R&D as a % of GDP
High Income Group Mean (19 countries)	3346	1.53
Medium Income Group Mean (17 countries)	104	1.02
Low Income Group Mean (17 countries)	35	0.48
Overall Mean (53 countries)	1244	1.02

Countries are classified as high income if their sample average GDP per capita exceeds $17000 (real 1995 dollars); low income if their sample average GDP per capita is below $3500 (real 1995 dollars); and medium if their incomes are in-between.

4 Empirical analysis

4.1 Case 1. Aggregate filings (by mode of filing)

The first forecasting exercises are with total EPO filings, aggregated across all technological fields. The total filings, however, can be broken down by mode of filing, depending on whether a particular filing is a first or a subsequent filing, and whether it is a direct EPO filing or an indirect one where the EPO is designated in a PCT application. Consider the following notation:

    EF   euro-direct first filings
    ES   euro-direct subsequent filings
    PF   Euro-PCT International phase first filings
    PS   Euro-PCT International phase subsequent filings
    Total= EF + ES + PF + PS

Table 7.2 shows some sample statistics on total filings as well as the composition of those filings: namely EF, ES, PF, and PS. To conserve space and to highlight the key stylized facts, only three years are shown: 1985, 1995, and 2000. The countries are grouped by bloc: EPC contracting states, Japan, US, and Other. Note that, because the data set goes up to 2000, the more recently joined contracting states of the EPC, such as Hungary and Romania, are treated as Other Bloc countries.

For EPC contracting states, a significant increase in euro-direct first filings has occurred as a share of all modes of filing. While a slight decline has occurred in the share of euro-direct filings that are subsequent filings (i.e. ES), there has been a greater increase in the share euro-PCT-IP filings that are subsequent filings (i.e. PS). For Japan and the US, direct first filings constitute a small share of filings at the EPO. Moreover, the share of EFs by the US and Japan has declined over time. Instead a tremendous increase in euro-PCT-IP filings has occurred, particularly subsequent filings (i.e. PS). In other words, PS is the most popular mode of filing for Japanese and US patent applicants.

Table 7.2. Breakdown of EP filings by route (euro-direct vs. euro-PCT-IP) and type of filing (first or subsequent)

Bloc	Year	%EF	%ES	%PF	%PS
EPC	1985	4.7	83.0	0.6	11.7
EPC	1995	8.6	48.7	1.5	41.2
EPC	2000	13.2	33.6	1.3	51.9
Japan	1985	3.4	87.5	1.2	7.9
Japan	1995	2.2	76.5	2.7	18.7
Japan	2000	1.2	60.1	3.2	35.5
USA	1985	4.5	74.2	2.1	19.2
USA	1995	1.9	35.4	3.3	59.4
USA	2000	2.6	20.3	1.6	75.4
Other	1985	7.7	55.7	3.2	33.5
Other	1995	4.2	18.8	7.2	69.8
Other	2000	1.5	14.3	5.4	78.8

EF, ES are euro-direct first filings and subsequent filings respectively. PF, PS are euro-PCT-IP first filings and subsequent filings respectively.

Among the Other bloc countries, there is quite a bit of variance in modes of filing. During the early 1980s, some of these countries had a relatively high share of euro-direct first filings in their EPO filings; however, by 2000, these countries have switched to using the PCT system to obtain patent protection in the EPO. Other bloc countries tend mostly to file subsequent patent applications to the EPO (whether directly or via the PCT).

Table 7.3. Filings by route: Total

Dependent variable: ln (TOTAL/Labor)			
	(1)	(2)	(3)
Constant	-0.109	-1.002***	0.017***
	(0.071)	(0.250)	(0.006)
Lag 1	0.694***	0.653***	0.469***
	(0.052)	(0.056)	(0.059)
Lag 2	0.123**	0.021	-0.075
	(0.065)	(0.069)	(0.064)
Lag 3	0.159***	0.256***	0.194***
	(0.072)	(0.049)	(0.056)
ln RD		0.073***	0.526***
		(0.020)	(0.144)
No. of Obs.	371	331	295
Adj. R-sq	0.98	0.98	
M2 p value			0.806
Method of Estimation	OLS	GLS	GMM

The equations are estimated over the period 1980–1996. OLS denotes ordinary least squares, GLS generalized least squares, and GMM generalized method of moments. Lag n refers to the dependent variable lagged n period(s), and ***, **, * denote significance levels of 1%, 5%, and 10% respectively. Standard errors are shown in parentheses. M2 p-value denotes the p-value associated with the test for 2nd order autocorrelation in the residuals. TOTAL refers to the sum of EF (Euro-Direct First Filings), ES (Euro-Direct Subsequent Filings), PF (euro-PCT-IP First Filings), and PS (euro-PCT-IP Subsequent Filings). ln RD is the natural log of gross expenditures on research and development (GERD) per worker (in real 1995 US dollars).

4.1.1 Estimation

Estimates of the forecasting equations are shown in Tables 7.3 and 7.4. In Table 7.3, the dependent variable is the natural log of total EPO filings per source country worker. In Table 7.4, the dependent variable is the natural log of EPO filings per worker by mode of filing (i.e. EF, ES, PF, and PS). Different model representations were examined (among others): AR3 (autoregressive model of order 3) and RE3 (regression model with the first three lagged values of the dependent variable and the flow of real R&D expenditures per source country worker). In other words, RE3 is AR3 augmented with R&D. Other variables were included in preliminary analyses, such as GDP per capita, but were found to be correlated with R&D and/or contributed marginally to the forecasting exercises (such as patent filing costs).

Thus the RE3 model is:

$$p_{it} = \eta_0 + \eta_1 p_{it-1} + \eta_2 p_{it-2} + \eta_3 p_{it-3} + \eta_4 r_{it} + \varepsilon_{it} \qquad (4)$$

where the lowercase letter p denotes the natural log patent filings at the EPO by source country i per source country labor, and r the natural log of source country research and development expenditures (in real 1995 US dollars) per source country worker. The AR3 model is where η_4 is set to zero.

The motivation for these different models is to highlight the role of R&D in predicting patent filings. A comparison, for example, between AR3 and RE3 shows whether R&D has any predictive power over and above the autoregressive model. A large fraction of the variation in the data can be captured by autoregressive terms (i.e. by the lags of the dependent variable) without any additional variables like R&D.[5]

In Table 7.3, the models were estimated over the period 1980–1996, so that out of sample forecasts for 1997–2000 can be made. Column 1 presents the results of the AR3 model. The coefficient on the first lag is just under 0.7. The patent-elasticity of R&D is measured to be 0.073 (meaning that a 1% increase in a source country's R&D leads, on average, to a 0.073% increase in its EPO patent filings) – if the equation is estimated by generalized least squares (random effects). If the model is estimated by generalized method of moments, the measured elasticity rises to 0.526. The R&D variable is strongly statistically significant by either method of estimation. For the GLS estimation, the null hypothesis of no correlation between the individual error term and the regressors could not be rejected. For GMM, the null hypothesis of no second-order autocorrelation in the differenced residuals cannot be rejected (which would otherwise indicate that the estimates are inconsistent).

In Table 7.4, for considerations of space, the AR3 results are not shown. Just the estimates of Eq. (4) by GLS and GMM are shown for each mode of filing. As with the case for total filings, GMM measures a higher elasticity of R&D. For EF filings (i.e. euro-direct first filings), none of the lagged dependent variables are statistically significant when estimated by

---

[5] The lag length was determined in preliminary examinations via quasi-likelihood ratio (QLR) tests; see Woolridge (2002), pp. 370-371.

GMM. Under GLS, all those variables are significant at or beyond conventional levels. As for ES filings (i.e. euro-direct subsequent filings), R&D is statistically significant at the 5% level under GLS but at the 10% level under GMM. As for the PF filings (i.e. euro-PCT-IP filings), R&D is statistically significant only at the 10% level under GLS but at the 1% level under GMM. Moreover, the second and third lagged dependent variables are insignificant under GMM. As for PS (i.e. euro-PCT-IP subsequent filings), R&D is measured to be important only under GMM.

Table 7.4. Filings by route: euro-direct, euro-PCT-IP, first and subsequent filings

Dependent Variable: ln (x/Labor)				
where x =	EF (1)	ES (3)	PF (5)	PS (7)
Constant	-1.142***	-0.418***	-2.191***	-0.802***
	(0.434)	(0.207)	(0.586)	(0.298)
Lag 1	0.536***	0.656***	0.548***	0.863***
	(0.056)	(0.043)	(0.058)	(0.043)
Lag 2	0.296***	0.214***	0.106*	0.038
	(0.061)	(0.051)	(0.062)	(0.049)
Lag 3	0.106**	0.108***	0.198***	0.039
	(0.056)	(0.043)	(0.053)	(0.034)
ln RD	0.071***	0.032**	0.07*	0.063
	(0.028)	(0.016)	(0.04)	(0.023)
No. of Obs.	330	460	247	353
Adj. R-sq	0.93	0.97	0.81	0.97
Method of Estimation	GLS	GLS	GLS	GLS
Dependent Variable: ln (x/Labor)				
where x =	EF (1)	ES (2)	PF (6)	PS (8)
Constant	0.019**	-0.043***	0.087***	0.044***
	(0.009)	(0.006)	(0.014)	(0.010)
Lag 1	-0.015	0.279***	0.234***	0.695***
	(0.063)	(0.046)	(0.065)	(0.052)
Lag 2	0.039	0.149***	-0.037	0.037
	(0.056)	(0.044)	(0.057)	(0.049)
Lag 3	0.044	0.229***	0.082	0.025
	(0.052)	(0.041)	(0.056)	(0.040)
ln RD	0.741***	0.268*	0.437***	0.343***
	(0.261)	(0.159)	(0.224)	(0.185)
No. of Obs.	292	413	218	308
M2 p-value	0.96	0.32	0.22	0.49
Method of Estimation	GMM	GMM	GMM	GMM

For terminology, see Table 7.3.

Thus, to sum up, the estimation results are sensitive to the method of estimation and to mode of filing. In particular, the effect of R&D (quantitatively and qualitatively) varies across different settings.

4.1.2 Forecast accuracy

Each of the forecasting equations discussed above can be compared for their ability to forecast. As each of the above equations was estimated over the period 1980–1996, the estimated – or fitted – equations can thus be used to generate out-of-sample forecasts of filings in 1997–2000. Table 7.5 reports the root mean square proportion error (RMSPE) associated with each equation's performance in predicting patent filings (whether it be total, EF, ES, PF, or PS) in year 1997, 1998, 1999, and 2000. Thus for each year, each of the models generates a forecast for each source country. By comparing these forecasts with the actual source country filings, the RMSPE computes a summary measure of the forecast errors across the source countries.

In general, it is tough to beat the AR3 model, but adding R&D does in some cases help to improve forecast accuracy (with some exceptions). For total filings, the model estimated by GLS tends to be best in the short run (1997 and 1998). AR3 has the lowest RMSPE for 1999 and GMM for 2000. The model estimated by GMM does best in the short run for predicting EF and PF. Otherwise, for euro-direct and euro-PCT-IP (first or subsequent) filings, the random effects model tends to do best for 1999 and 2000.

4.1.3 Sample forecasts

Table 7.6 provides some sample forecasts for a select sample of source countries. In this table, forecasts for total EPO filings in 1998 are used as an example. This could easily be replicated or reproduced for other years and for detailed modes of filing: EF, ES, PF, and PS.

For total filings, the AR3 and RE3 both under-predict the overall (i.e. all country) filings, but not by much. The shortfall is about 7% of the actual filings. For the US, the RE3 model's forecast is 97% of actual; for Japan, it is about 95%, and Germany 86%. For overall EF, the AR3 and RE3 models are off by about 1 800 filings (which are about a quarter of actual filings). The model estimated by GMM produces a sum forecast which is (marginally) closest to the actual. In the bigger scheme of things, the total forecasts and forecasts by source country are mildly different across the different models and methods of estimation.

Table 7.5. Summary of forecast errors: Aggregate EP filings

Model	Variable	N	Root Mean Square Proportion Error			
			1997	1998	1999	2000
AR3	Total	371	0.153	0.149	0.125	0.190
RE3	Total	331	0.139	0.145	0.126	0.189
GM3	Total	295	0.159	0.167	0.163	0.173
AR3	EF	362	0.378	0.545	0.359	0.422
RE3	EF	330	0.379	0.539	0.341	0.455
GM3	EF	292	0.276	0.405	0.474	0.517
AR3	ES	511	0.298	0.319	0.502	0.278
RE3	ES	460	0.307	0.315	0.483	0.273
GM3	ES	413	0.378	0.420	0.604	0.367
AR3	PF	249	0.473	0.424	0.418	0.759
RE3	PF	247	0.409	0.341	0.383	0.719
GM3	PF	218	0.394	0.217	0.447	1.081
AR3	PS	393	0.293	0.289	0.175	0.275
RE3	PS	353	0.245	0.242	0.170	0.225
GM3	PS	308	0.372	0.271	0.270	0.218

Notes: The forecast errors correspond to the estimated models in Tables 7.3 and 7.4 (but the AR3 results were omitted in Table 3B for space considerations). Each entry is the RMSPE (root mean square proportion error) associated with each model. AR3 autoregressive model of order 3, RE3 random effects model of R&D and three lags of the dependent variable, and GM3 model of R&D and three lags of the dependent variable estimated by GMM. N denotes the number of observations, EF euro-direct first filings, ES euro-direct subsequent filings, PF euro-PCT-IP first filings, and PS euro-PCT-IP subsequent filings. Total is the sum (=EF+ES+PF+PS).

Table 7.6. Sample forecasts of EP filings: Actual vs. predicted values for 1998

	Total Filings:			
Source Country	Actual	AR3	RE3	GM3
France	7115	6671	6653	6164
Germany	19860	17557	16971	17381
Japan	16169	15395	15479	15784
UK	5041	5449	5370	4969
USA	36860	34918	35548	35868
Other	25378	22701	22168	23086
All 53 Countries	110423	102691	102189	103251

AR3 refers to the autoregressive model of order 3, RE3 to the random effects model with 3 lags of the dependent variable, and GM3 to the model with 3 lags of the dependent variable estimated by GMM.

4.2 Case 2. Sectoral filings (joint clusters)

EP filings can be broken down by technological field. The patent applications at the EPO are put into one of fourteen technological divisions, called joint clusters (JC):

- JC1   Electricity and electrical machines
- JC2   Handling and processing
- JC3   Industrial chemistry
- JC4   Measuring and optics
- JC5   Computers
- JC6   Human necessities
- JC7   Pure and applied organic chemistry
- JC8   Audio, video and media
- JC9   Civil engineering and thermodynamics
- JC10   Electronics
- JC11   Polymers
- JC12   Biotechnology
- JC13   Telecommunications
- JC14   Vehicles and general technology

Fig. 7.1 shows the percentage distribution of filings by joint cluster (JCs 1–14) for 1998. Most of the EPO filings were in the field of handling and processing, followed by pure and applied organic chemistry and human necessities. Relatively small shares of filings occur in the new (emerging) fields of computers, biotechnology, and telecommunications.

Instead of analyzing the filings in each of these JC technological fields, this chapter examines functionally related groups of joint clusters. A judgment call has to be made as to which JC classes should be grouped together. It is beyond the scope of this chapter to determine the best grouping. Rather the objective is to apply forecasting methods to paneled groups of JC filings. Thus the fourteen JC fields are put into the following five technology groups (G1 to G5):

G1	Group 1	Electricals	JCs 1, 8, and 10
G2	Group 2	Chemicals	JCs 3, 7, 11, and 12
G3	Group 3	Manufacturing	JCs 2, 6, and 14
G4	Group 4	Physics	JCs 4 and 9
G5	Group 5	Computer related	JCs 5 and 13

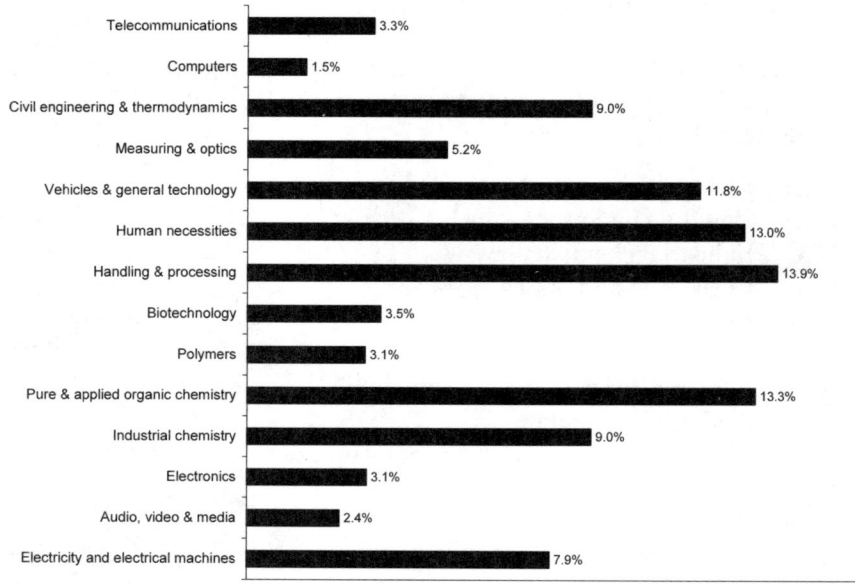

Fig. 7.1. Percentage of EP Filings by joint cluster: 1998

4.2.1 Estimation

Tables 7.7 and 7.8 show estimates of the model by technology group (where Table 7.7 presents the GLS results and Table 7.8 the GMM results). For reasons of space, only the total filings (that is, the sum of EF, ES, PF, and PS) are shown. Also the results for the AR3 model are omitted.

The models are estimated from 1980–1996, so that out-of-sample forecasts can be generated for 1997–2000. First, the GLS results suggest that, other than for Group 1 (Electricals) and Group 5 (Computer related), only the first lag of the dependent variable explains patent filings. Business enterprise R&D is a statistically significant determinant of total filings for all groups. Under GMM estimation, the measured elasticity of R&D is much higher, but nonetheless qualitatively significant at conventional levels. Moreover, under GMM estimation, typically the second and third lags of the dependent variable are statistically insignificant (except in the case of Group 2), and the coefficients of the first lag are under 0.3 indicating a low degree of persistence or momentum in filings. For Group 5 filings, however, the null of no second-order autocorrelation can be rejected, suggesting that estimates of the model are not consistent.

Table 7.7. Sectoral regressions on total filings by generalized least squares

Dependent Variable: ln (TOTAL/Labor)
By Technology Group:

	G1	G2	G3	G4	G5
Constant	-0.054	-0.129	-0.149	-0.441***	-0.269
	(0.235)	(0.204)	(0.145)	(0.177)	(0.387)
Lag 1	0.586***	0.670***	0.955***	0.717***	0.506***
	(0.074)	(0.073)	(0.068)	(0.073)	(0.075)
Lag 2	0.141*	0.131	-0.089	0.139*	0.283***
	(0.079)	(0.086)	(0.086)	(0.085)	(0.080)
Lag 3	0.146**	0.078	0.081	0.051	0.076
	(0.067)	(0.065)	(0.065)	(0.063)	(0.077)
Business-Enter. R&D	0.163***	0.135***	0.042**	0.063***	0.160***
	(0.052)	(0.034)	(0.020)	(0.024)	(0.074)
No. of Obs.	175	176	176	176	173
Adj. R-sq	0.94	0.96	0.97	0.96	0.86
Method of Estimation	GLS	GLS	GLS	GLS	GLS

Business Enter. R&D is the natural log of Business Enterprise Research and Development (BERD) expenditures per worker. For other terminology, see Table 7.3.

Table 7.8. Sectoral regressions on total filings by generalized method of moments

Dependent Variable: ln (TOTAL/Labor)
By Technology Group:

	G1	G2	G3	G4	G5
Constant	0.015*	0.011**	0.011	0.016***	0.067***
	(0.008)	(0.004)	(0.004)	(0.004)	(0.015)
Lag 1	0.266***	0.207***	0.283***	0.224***	0.125*
	(0.075)	(0.071)	(0.063)	(0.069)	(0.076)
Lag 2	0.044	0.136**	0.053	0.089	0.048
	(0.071)	(0.063)	(0.059)	(0.065)	(0.077)
Lag 3	0.074	0.040	-0.018	-0.072	-0.009
	(0.069)	(0.054)	(0.051)	(0.059)	(0.087)
Business-Enter. R&D	0.768***	0.642***	0.652***	0.627**	0.868***
	(0.132)	(0.072)	(0.067)	(0.081)	(0.205)
No. of Obs.	156	156	156	156	154
M2 p-value	0.25	0.39	0.38	0.38	0.02
Method of Estimation	GMM	GMM	GMM	GMM	GMM

For terminology, see Tables 7.3 and 7.7.

Table 7.9. Summary of forecast errors: Sectoral filings (by joint cluster)

G1 Technology Group 1 (Electricals):						
Model	Variable	N	1997	1998	1999	2000
AR3	Total	177	0.205	0.155	0.111	0.248
RE3	Total	175	0.198	0.166	0.113	0.199
GM3	Total	156	0.126	0.147	0.111	0.194

G2 Technology Group 2 (Chemicals):						
Model	Variable	N	1997	1998	1999	2000
AR3	Total	178	0.162	0.193	0.113	0.092
RE3	Total	176	0.154	0.183	0.107	0.093
GM3	Total	156	0.123	0.189	0.114	0.117

G3 Technology Group 3 (Manufacturing):						
Model	Variable	N	1997	1998	1999	2000
AR3	Total	178	0.123	0.072	0.107	0.110
RE3	Total	176	0.118	0.073	0.095	0.096
GM3	Total	156	0.101	0.128	0.116	0.137

G4 Technology Group 4 (Physics):						
Model	Variable	N	1997	1998	1999	2000
AR3	Total	178	0.155	0.129	0.116	0.135
RE3	Total	176	0.149	0.130	0.117	0.114
GM3	Total	156	0.092	0.143	0.139	0.141

G5 Technology Group 5 (Computer related):						
Model	Variable	N	1997	1998	1999	2000
AR3	Total	175	0.345	0.285	0.267	0.306
RE3	Total	173	0.325	0.298	0.255	0.304
GM3	Total	154	0.224	0.271	0.186	0.221

The forecast errors correspond to the estimated models in Table 6A–B. Each sub-table here represents a different technology group. Each entry in each sub-table is the RMSPE (root mean square proportion error) associated with a model: AR3 autoregressive model of order 3 (whose estimation results are omitted in Table 7 to conserve space), RE3 random effects model of R&D and three lags of the dependent variable, and GM3 the regression model of R&D and three lags of the dependent variable estimated by GMM. N denotes the number of observations and Total the sum of all filings, aggregated across different routes (i.e. TOTAL=EF+ES+PF +PS, where EF is euro-direct first filings, ES euro-direct subsequent filings, PF euro-PCT-IP first filings, and PS euro-PCT-IP subsequent filings).

4.2.2 Forecast accuracy

Table 7.9 reports on the root mean square proportion errors associated with each of the technology group regressions discussed above. Each of the estimated models was used to make forecasts for 1997 – 2000 inclusive. Note that panels AE of Table 7.9 refer to Groups 1 to 5 respectively.

In general, the forecast performance is improved using R&D (relative to that of the AR3 model), whether the model is estimated by GMM or GLS. For predicting Group 1 filings, the model with R&D produces lower forecast errors than the AR3, although for 1999, the GLS estimations produce a slightly higher forecast error. For Group 2 filings, the model with R&D performs better than the AR3 for the very short run (1997 and 1998). For 1999 and 2000, the model estimated by GLS performs best in relative terms. For predicting Group 3 and 4 filings, the model estimated by GMM produces relatively the largest errors from 1998 on. The model estimated by GMM produces relatively smaller errors for 1997 and 2000. There is not much improvement over AR3 for 1998–1999. Finally, for predicting Group 5 filings (Computer related), the model with R&D as estimated by GMM produces relatively the lowest forecast errors. Thus forecasting can be enhanced using R&D as a predictor, but there is no definite forecasting advantage exhibited by either method of estimation.

4.2.3 Sample forecasts

Next, the forecast totals for 1998, summed across individual source countries, can be seen in Table 7.10.

For total filings, the model does quite well for technology groups 1–4. For instance, in terms of overall country filings, the predicted sum of Group 1 filings is 90% of the actual sum. The predicted cross-country sum of Group 2 filings is 94.3% of the actual sum; for Group 3 it is 95.8%, and for Group 4, it is 91.3%. But for Group 5, the predicted overall sum of filings is 74.1% of actual. Quantitatively, there are not as many filings in Group 5 as there are in each of the other groups. Perhaps a greater degree of uncertainty or unpredictability is a characteristic of innovation in computers and telecommunications such that actual patenting activity deviates substantially from what trends in R&D and past patenting behavior would suggest.

Table 7.10. Sample filing forecasts by technology groups: Actual vs. predicted Values for 1998

Technology Group	Actual	RE3 Model	GM3 Model
G1	19388	17479	18093
G2	31860	30269	29344
G3	28366	27084	25724
G4	13890	12302	12207
G5	10795	8001	8836

The model estimated by GMM produces a predicted sum of filings (aggregated across source countries) that is somewhat closer to the actual filings for Groups 1, 3, and 5.

4.3 Case 3. Patent family filings

In this section, attention is shifted from patent applications to patent families. International patent applications and other documents relating to the same invention would comprise a patent family. The patent family data are indexed by the priority number of the first filing, with information on subsequent filings for that invention in four blocs (EPC contracting states including the EPO, US, Japan, and Other countries). The database can be filtered to select different types of patent families (e.g. trilateral patent families, which are families that involve patent filing activity in each of the trilateral blocs: US, Japan, and EPC).

Patent family data thus depend on the appearance of patent publications that can index a patent family (link subsequent filings and priority filings). There is a timeliness problem due to this dependence on patent publication lags. Consequently, there may be some under-reporting of subsequent filing activities connected with an earlier priority forming filing. This particularly affects the more recent years, such as 1999 and on. The timeliness problem particularly affected data for the US where, until 2000, patents were published only upon grant. Thus the figures for the US could only be updated after several years of delay.

Table 7.11 provides a breakdown of international patent families that are based on priorities filed in 1999. The format follows that of the Statistical Annex of the Trilateral Statistical Report 2004 edition (EPO, JPO, USPTO 2005.) The 1999 figures are provisional, so that there will be an underestimation of patent family formation for this particular year. In this table, the subsequent filing destinations are available only by blocs (EPC, US, Japan, and Other). The first column provides the quantity of first filings associated with the priority country and the remaining columns show different types of subsequent filings as a percentage of first filings. Note that these percentages (in each row) need not sum to 100% since the various bloc combinations shown are not mutually exclusive.

Table 7.11. Patent families derived from first filings in 1999, by country of origin

Priority Year 1999		Families with activity outside the Bloc of Origin (% Priority filings claimed in Country of Origin From):						
Country Bloc	First Filings	All Other Blocs	Other Trilateral Blocs	EPC	Japan	USA	Other Countries	Trilateral Patent Families
EPC	130999	34.1%	17.1%	-	8.3%	12.3%	24.4%	3.9%
Japan	356397	12.7%	11.6%	8.3%	-	7.1%	5.1%	3.8%
US	153350	41.6%	30.1%	28.7%	9.2%	-	32.2%	7.8%
Other	157888	6.2%	6.2%	3.5%	2.1%	2.5%	5.1%	0.5%

Source: European Patent Office, PRI database.

In 1999 (as in all other years), Japan leads with the most first filings (in excess of 350 000), the vast majority of which are domestic filings. The very large number of such filings has been attributed to the practice in Japan of filing domestic patent applications with single claims. The US has the second most priority filings, with a total of 153 350, which is slightly more than the Other Country Bloc. The EPC as a whole has 130 999 first filings, almost half of which come from Germany. The next most productive country for first filings within the EPC is the UK, followed by France. About 12 000 euro-direct first filings at the EPO took place in 1999, which are just under 10% of all first filings within the EPC bloc. Among the Other Country Bloc, most first filings occur in China, followed by Korea, Russia, and Australia.

More than 40% of US first filings formed patent families with at least one other bloc (EPC, Japan, or Other), 30.1% formed patent families with at least one other trilateral bloc (Japan, EPC, or both), 32.2% formed patent families with at least a non-trilateral country or bloc, 28.7% formed patent families with at least the EPC, 9.2% with at least Japan, and just under 8% formed a trilateral patent family. For Japan, the percentages of patent family formation are generally smaller due to the large number of its first filings (i.e. in the denominator). For the EPC as a whole, the rates of patent family formation are generally between those of the US and Japan. Switzerland leads with the highest rate of trilateral patent family formation, followed by France, then Germany. The medium to lower income EPC states, such as Greece, Ireland, Monaco, Portugal, and Spain produce a negligible number of trilateral patent families. Among the Other Bloc countries, the largest number of trilateral patent family counts comes from Korea, and very small or zero trilateral patent families come from Brazil,

Bulgaria, China, Hungary, Mexico, Philippines, Poland, Romania, and Russia[6]. The EPC obtains relatively most secondary filings from Korea, Australia, Canada, New Zealand, Norway, South Africa, and Israel. Again, it should be remembered that the observations for year 1999 should be treated as provisional until further subsequent filing data are obtained.

4.3.1 Estimation

Patent family formation rates are now the dependent variables of interest. The focus is on the numbers of families with activity in the EPO, other countries, other trilateral blocs, or trilateral patent families.

In the case of patent family formation, lagged priority filings are used as a regressor (which will replace the second and third lags of the dependent variable, since lagged priority filings capture past patenting and innovative activity). Conceptually, the priority filings reflect the overall level of new inventions that are patented in a particular period; the subsequent filings measure the transfer of those patentable inventions abroad. Thus priority filings are a measure of inventiveness, subject to the qualification that not all inventions are patented or are patentable, while subsequent filings are a measure of international technology diffusion. Here, first filings refer to priority filings of the source country, where source country refers to the country of origin of the first filings (from which priority claims are made by all other patents) and not necessarily to the country of residence of the patent owner or inventor.

R&D is modeled as a determinant of subsequent filings. That is, R&D affects the innovative potential of a source country and the propensity to make subsequent filings. One reason that research and development can stimulate the transfer of technologies abroad is that the R&D expenditures may reflect the investment effort level (and possibly thereby the quality level) of a source nation's patentable inventions. Thus, the greater a source country's resources devoted to R&D per worker, the greater the number of innovations that might be worthy of patenting subsequently in other markets. Support for this view comes from previous studies on patent valuation. This research indicates that worthy patents can be "screened" by observing which ones are renewed frequently over time or which ones are used to apply for patents in more destinations (markets), and thereby form

---

[6] Note that for the Other Bloc, the second and third columns in Table 7.11 coincide as a result of the fact that where a patent family is formed with at least one other bloc, that bloc is one of the trilateral blocs.

larger families.[7] Thus, on theoretical grounds, R&D could stimulate both first and subsequent filings. The greater the R&D content, for example, in a nations' supply of patentable inventions, the greater the proportion of priority filings that will be likely used as a basis for subsequent filings in other markets.

In what follows, models of subsequent filing behaviour are estimated for the sample years 1981–1994, corresponding to the priority filing years 1980–1993. The estimated models are then used to generate out-of-sample forecasts for families containing subsequent filings in 1995–2000. We then compare these forecasts with actual data. It will be seen that the models generally over predict for 1999 and 2000, which is consistent with the fact that there is a timeliness problem in reporting families incorporating subsequent filing activity for these years. It is for this reason that the out-of-sample forecast interval was expanded and in-sample estimation period decreased.

Table 7.12. Patent family regressions: Families with activity in the EPO and trilateral patent families

Dependent Variable: ln (Patent families involving x / Labor)				
where x =	EPO (1)	EPO (2)	Trilateral (3)	Trilateral (4)
Constant	-2.705***	0.039***	-3.238***	0.025***
	((0.427)	(0.007)	(0.520)	(0.007)
Lag 1	0.829***	0.348***	0.828***	0.224***
	(0.021)	(0.055)	(0.026)	(0.056)
ln FF	0.062***	0.154**	0.037*	0.241***
	(0.021)	(0.065)	(0.023)	(0.063)
ln RD	0.250***	0.083	0.279***	0.574***
	(0.036)	(0.132)	(0.043)	(0.167)
No. of Obs.	469	425	429	384
Adj. R-sq	0.97		0.97	
M2 p-value		0.03		0.03
Method of Estimation	GLS	GMM	GLS	GMM

FF refers to priority forming first filings. For other terminology, see Table 7.3.

---

[7] For studies that infer patent value from patent renewal behaviour, see Pakes (1986), Schankerman and Pakes (1986), and Lanjouw et al. (1998). For studies that infer patent value from patent family size, see Harhoff et. al. (2003), Lanjouw and Schankerman (1999), and Putnam (1996). Harhoff et. al. (2003), p. 1343, for example, argue that "patents representing large international patent families are especially valuable."

The estimation results are in Table 7.12. For each type of dependent variable, GLS and GMM estimates are provided. The first two columns focus on counts of families with activity at the EPO and the last two columns on trilateral patent families. For each type of dependent variable, an AR3 model was estimated, though the results are not shown. Thus, in effect the comparison is between the forecast ability of an AR3 versus that of a model with a lagged dependent variable, lagged priority filings, and R&D. Note that first filings are lagged one period because, in accordance with the Paris Convention as discussed in Sect. 1, applicants have up to one year to file subsequent patent applications and to refer to the priority date associated with the first filing.

Random effects estimation indicates that a 1% increase in priority filings leads to a 0.062% growth in families with activity at the EPO, whereas a 1% increase in R&D stimulates a 0.25% increase in those filings. For trilateral patent families, a qualitatively similar pattern is exhibited. The previous period's patenting activity or intensity influences a current period's activity. R&D also has a statistically significant influence on the technology transfer rates. However, lagged priority filings weakly determine the number of trilateral patent families. This finding implies that the mere size or stock of patentable inventions (as measured by the flow of priority filings) does not influence international technology diffusion (holding other factors constant). It is, in other words, not necessarily the case that the more inventions there are the more international patent family formation. Some countries have a relatively large number of priority filings, yet have a comparatively smaller propensity to patent abroad (e.g. Japan), while others have a relatively small number of priority filings, yet have a comparatively high propensity to patent internationally (such as Canada). This result might suggest that patentees make separate decisions concerning their priority filings and subsequent filings.

GMM estimates paint a somewhat different picture. The coefficients of lagged first filings are statistically significant at conventional levels. The measured elasticity of R&D is generally higher, but R&D is found to be statistically insignificant in explaining families with activity at the EPO. However, the presence of second-order autocorrelation suggests that the estimates are not consistent.

4.3.2 Forecast accuracy

Table 7.13 shows the root mean square proportion errors (RMSPE) associated with each patent family model estimated thus far. Since the estimated equations in Table 7.12 were estimated up to year 1994, the actual (realized) values of the independent variables for 1995–2000 are used to make predictions of the dependent variable.

The forecast performance between AR3 and RE1 (i.e. model estimated via random effects) is fairly similar for families with activity at the EPO. The AR3 performs relatively better for the earlier years (1995–1997) but the momentum captured in the AR3 does not extend well into a longer time horizon. For the number of trilateral patent families, the AR3 performs generally better than the RE1 model (which incorporates lagged priority filings and R&D per worker). However, the RE1 model generally performs better than the simple AR1 model. Nonetheless the differences in forecast errors between AR3 and RE1 are of small magnitude. Note the relatively large forecast errors for predicting trilateral patent family filings for 1999 and 2000. This reflects the timeliness problem of obtaining actual or realized trilateral patent family data for those years.

For predicting families with activity at the EPO, the model estimated by GMM produces larger forecast errors. (Recall that in this case, R&D does not have explanatory power, but that the estimates are not consistent.) Likewise, for predicting trilateral patent families, GMM produces larger forecast errors (except for 1997) and gives estimates that are not consistent (see Table 7.13). Thus, in general, AR3 and random effects estimation perform better on the RMSPE criterion.

Table 7.13. Summary of forecast errors: Patent family regressions

			Root Mean Square Proportion Error		
Model	Variable	N	1995	1998	2000
AR3	EPO	405	0.319	0.257	1.330
RE1	EPO	469	0.420	0.250	1.312
GM1	EPO	425	0.731	0.278	1.392
AR3	Trilateral	366	0.414	0.812	0.850
RE1	Trilateral	429	0.530	0.859	0.854
GM1	Trilateral	384	1.104	0.988	1.924

4.3.3 Sample forecasts

Table 7.14 provides some sample forecasts by source country for 1998. For predicting the number of patent families with activity at the EPO, the RE1 model does generally better overall. The predictions are quite close for US, Japan, and Germany. The RE1 predicts a total, country-wide, forecast of 102 150 families with subsequent EPO filings. This is just 158 filings shy of the actual (or an error rate of just 0.154%).

Table 7.14. Sample patent family forecasts: Actual vs. predicted values for 1998

	(A) Families with activity at the EPO:			
Source Country:	Actual	AR3	RE1	GM1
France	6087	5983	5809	6241
Germany	20088	18558	17314	20463
Japan	21000	20428	23258	21367
UK	6151	5748	5106	6252
USA	35627	32785	37152	33768
Other	13355	13336	13511	13727
All 53 Countries	102308	96838	102150	101818
	(B) Trilateral Patent Families:			
Source Country:	Actual	AR3	RE1	GM1
France	1537	5983	5809	6241
Germany	5267	18558	17314	20463
Japan	17119	20428	23258	21367
UK	1549	5748	5106	6252
USA	24058	32785	37152	33768
Other	3338	13336	13511	13727
All 53 Countries	102308	96838	102150	101818

(A) represents families with activity at the EPO by the source country, (B) represents trilateral patent families by the source country. AR3 are forecasts from an autoregressive model of order 3, RE1 are forecasts from a random effects model of R&D and a lagged dependent variable, and GM1 are forecasts from a model of R&D and a lagged dependent variable estimated by GMM.

For predicting the formation of trilateral patent families, all the models over-predict. While the total, country-wide, forecasts of trilateral patent families are relatively not as good, the forecasts for individual countries vary in accuracy. For example, for US, the RE1 model produces good forecasts. But generally, the model over-predicts these kinds of filings for most countries. The errors appear to be systematic and suggest that forecasts of 1998 may also suffer from the timeliness problem. Clearly, the model has fit well historically over the truncated sample period (up to 1994), yet the RMSPE for 1998 exceeds 0.8 (suggesting that the predictions are 80% greater than, or almost double, the actual). This means either that the model has omitted important variables with predictive content, or that there is a severe lag in the reporting of actual trilateral patent families. It would be useful to re-do the forecasts at a later point in time in order to determine the more likely source of the forecast errors.

The model estimated by GMM (i.e. GM1) also provides a total forecast of families with activity at the EPO that is quite close to the actual, but GM1 overpredicts trilateral patent families. However, country by country, the pattern of forecasts appears to be qualitatively similar, independently of the way the model is estimated. Although more testing is desired, obtaining good (practical) forecasts may not be too sensitive to the underlying method of estimation.

5 Conclusions

Patenting activity is intense within the EPO. The US, Japan, and the EPC contracting states together account for the bulk of world patenting activities. This is in large part due to their relatively high incomes (which provides for larger markets), their greater productivity (which makes their inventors more prolific producers of knowledge capital), their greater R&D and science and engineering resources, and their stronger intellectual property systems. But the increased patenting in the EPO is also attributable to the institutional system itself. With the benefits of single filing and centralized procedures, the economic cost of patenting in EPO member countries has been reduced. Moreover, membership in the EPO has particularly helped the smaller member economies to obtain increased technology inflows. This is largely due to the low marginal cost of designating additional EPO states (beyond the top three to five states).

By 1999, nearly all the EPO states have been designated in EPO patent applications. Despite reducing fees and improving procedures for filing patents, the EPO receives very few patents from developing countries or emerging markets. This is due in good measure to factors internal to those nations (their policies and environment). However, to the extent that they depend on access to foreign markets in Europe, US, and Japan for their development, their lack of involvement in international patenting activities becomes an important issue.

The main highlights of the empirical investigation are as follows:

- R&D better explains total filings rather than the different modes of filing (EF, ES, PF, and PS). It does not explain well why agents choose to file at the EPO first or subsequently or why they would (or would not) opt to seek protection in the EPO via the PCT system.
- Models with lags of the dependent variable (usually the first three lags) plus R&D produce quite good forecasts. Typically 90% or more of the actual filings is predicted. Note though that these forecasts are generated using actual R&D values realized during the forecasting period. In practice, for predicting filings beyond the sample period (e.g. 2005 and beyond), forecasts of R&D (as well as models of R&D behavior) need to be developed.
- Generally, predictions of euro-direct (or euro-PCT-IP) subsequent filings are better (in the sense of lower RMSPE) than predictions of euro-direct (or euro-PCT-IP) first filings.
- Forecasts by technology (joint cluster) groups were also quite close to actual, except for Group 5 (computers and telecommunications). As with forecasting aggregate filings (across technological fields), the models forecast better for total filings rather than filings by different modes. Also, at the sectoral level, subsequent filings are easier to predict than first filings.
- For predicting counts of patent families, R&D is useful for predicting families with activity at the EPO and trilateral patent families.
- Lagged priority filings were also useful at predicting numbers of families with activity at the EPO but weakly useful at predicting trilateral patent families. This is because, across countries, there is not a tight connection between domestic filings and international filings (since some countries have large domestic filings and relatively less international, and vice versa). Of course, the transfer rate (i.e. ratio of subsequent filings to first filings) may be stable over time for each country, but across countries there does not seem to be a monotonic relationship between the transfer rate and level or size of first filings.
- Due to the timeliness problem, forecasts of patent families were less accurate for 1999 and 2000 (and even for 1998 in the case of trilateral patent families). For periods where the timeliness problem is not a problem, the forecasts are generally quite good (i.e. close to 99% of actual).

In conclusion, there are several possible extensions to this study. Firstly, more research is needed to better understand and explain patent granting behavior. This requires more detailed knowledge about the patenting authorities – their objectives and constraints. The existing literature has predominantly focused on the behavior of patent applicants. Secondly, it would be of interest to explore the feedback effects, if any, from the EPO system to R&D activities in various source countries. Inventors or firms should perceive the EPO region to represent a large (or larger) market. This surely should have impacted on their incentives to do R&D (including inventors in developing countries). The larger market should justify a larger investment in innovation or greater risk-taking. Thirdly, it would also be useful to derive destination variables for the EPO. Thus far the models of patenting behavior in this study have only incorporated source country characteristics. The chapter discussed the challenges posed in constructing destination variables. Nonetheless, in future work, destination variables should be explicitly modeled.

Finally, the panel data framework above imposes the same coefficients (on say R&D) for all source countries. It would be useful in the future to model source country heterogeneity (in, for example, their patenting responses to R&D). Panel data are useful where the time-series dimension is not especially large (e.g. 1980–2000 for the full sample or 1980–1996 for the forecasting sample) so that cross-sectional observations add more variability. It would be valuable, though, to capture differences in "slopes" by source country.

Annex: Technical notes

The following are brief sketches on the methods of estimation used in this chapter. Interested readers are referred to Baltagi (2001) for more details. The statistical software package, STATA, provides routines to conduct these estimation methods on panel data[8]; for example, the commands xtregre performs generalized least squares (random effects) and xtabond performs generalized method of moments for models with lagged dependent variables. STATA also provides a number of post-estimation commands, such as predict, to generate out-of-sample forecasts.

- Generalized Least Squares (GLS)

Suppose $P_{it} = \beta_0 + \beta_2 x_{it} + \upsilon_i + \mu_{it}$, where P denotes the log of patent filings per worker and x an exogenous regressor (such as the log of research and development per worker). Under GLS, the individual effect υ_i is assumed to be random, and the variables are transformed. For example,

---

[8] See http://www.stata.org.

$$P_{it}' = P_{it} - \theta_i \overline{P}_i$$
$$x_{it}' = x_{it} - \theta_i \overline{x}_i$$

where

$$\overline{P}_i = \sum P_{it}/T_i, \ \overline{x}_i = \sum x_{it}/T_i, \text{ and } \theta_i = 1 - \sqrt{\frac{\text{var}(\hat{\mu}_{it})}{T_i \text{var}(\hat{v}_i) + \text{var}(\hat{\mu}_{it})}}$$

such that a least squares regression is run on the transformed variables. For consistent estimates, the random effects specification requires no correlation between x_{it} and v_i.

- Generalized Method of Moments (GMM)

With a lagged dependent variable, the model becomes:

$$P_{it} = \beta_0 + \beta_1 P_{it-1} + \beta_2 x_{it} + v_i + \mu_{it}$$

Correlation exists between v_i and one of the explanatory variables (namely the lagged dependent variable), since the above equation holds in the previous period and the individual effect is time-invariant (thus P_{it-1} is a function of v_i). GLS would be biased. The traditional way to solve this problem is to first difference and eliminate the individual effect and use lagged differences and lagged levels of variables as instruments. The Arellano and Bond (1991) method allows for the exploitation of more sample information; for example, the orthogonality conditions between the disturbances and the lagged values. For example, if the moment conditions are:

$$E[(Z^T(P - x\beta)] = 0$$

where Z is the matrix of instruments and ^{T} indicates transposition, GMM involves choosing the value of the parameters β to minimize the following loss function:

$$L = (P - x\beta)^T Z \hat{W} Z^T (P - x\beta)$$

where $\hat{W}$ is a symmetric, positive semi-definite (estimated) weighting matrix.

Chapter 8

Micro data for macro effects*

Rainer Frietsch

Innovation Systems and Policy, Fraunhofer Institute for Systems and Innovation Research, Karlsruhe, Germany

Modules: A, B.
Software used: Excel, SPSS, STATA, Word

1 Introduction

Forecasting future patent applications is an important task in many respects. For public policy and administration, it can be a kind of early warning system to intervene with the help of adapted funding behaviour, if the innovation system happens to move in an unintended direction. From a research perspective, forecasting can be used to give evidence of future competitiveness. For the European Patent Office (EPO) – as well as for any other patent office – it is first of all a tool to anticipate future workloads and to plan future budgets.

The chapters so far presented in this volume deal with the forecasting problem first of all on the macro-level of all applicants or on a meso-level of countries, technological fields or sectors. The aim of this chapter is to present two micro-approaches on the level of applicants that can be applied in the context of forecasting future patent filings. The first is a random sample of patenting firms, which was combined with data from an international firm database. The second micro database was set up by combining the so called R&D Scoreboard, published by the Department of Trade and Industry (DTI) of the British government, and the patent database at the EPO.

The intention of this study is to test the feasibility of combining the EPO patent database with some external data sources and to use this new data set to investigate the dependency of patent filings on structural and economic factors. The two micro approaches – random sample and sample of

---

* An earlier version of this chapter was presented at the WIPO-OECD Workshop on the Use of Patent Statistics, Geneva, 11th/12th October 2004.

R&D-intensive companies – allow for different kinds of perspectives. Whereas the first approach is able to cover the "real" distribution of applicants and their structure at the EPO, the second dataset may uncover further structural determinants of patent filings by an exclusive look at the largest applicants.

In most cases, surveys are used to analyse the dependencies of patents and other economic factors at the micro-level – either existing surveys or self-conducted surveys. Existing surveys like the Community Innovation Survey (CIS) or its national versions may provide the database for the analysis in question. These surveys have a broad focus on innovative companies and on the innovation system, but their usefulness for the analysis of patent applications is restricted, on the one hand, because only about 30% of the companies covered by the CIS use patents at all, and on the other hand, asking people about the number of patent applications always leads to some bias. Furthermore, it is intended to account for the total amount of patents or changes therein over time and not for the applications at one certain office. Another survey – focusing more on patent applications in Europe – is the Applicant Panel Survey that is conducted by the European Patent Office (EPO 2006). Although the immediate aim of this survey is to forecast future filings, it takes a technological perspective and is centred around R&D activities. A similar activity is undertaken by the USPTO, where they ask a stratified sample of large companies, SMEs, universities and free inventors about their future application intentions (see OFR 2004a, 2004b). The results of this activity do not seem to be for public use, as the information is very restricted in this respect. The JPO also surveys its applicants, with the intention of learning more about actual and future activities of their clients, using a very large sample (about 16 000 firms)[1]

The idea of linking patent data and economic information on applicants is not new. Many projects have been conducted in this respect. Among the most prominent ones is the NBER Patent Citation Database (Hall et al. 2001) which covers first of all US-American grants over a period of about 30 years. Several research institutes located in Europe also developed such datasets (for example OST, SPRU, INCENTIM etc.), but these are, on the one hand, not publicly accessible and, on the other hand, they often focus on national idiosyncrasies. None of these sources can directly be used for forecasting EPO filings.

All the mentioned databases have advantages as well as disadvantages when the question of forecasting future filings is addressed. There is still some need for further improvements of data for this kind of activities, es-

---

[1] The report is published in Japanese; see http://www.jpo.go.jp/.

pecially at the micro-level. Having this in mind, two alternative databases are suggested in this chapter. The intention was to come up with an approach that is easy to implement, not too time- and resource-consuming and therefore easy to replicate, maybe annually.

For the purposes and analyses presented in this chapter, any statistical software package could have been used. SPSS was applied for the descriptive statistics as this software (version 11.0 and higher)[2] offers the possibility to directly drag and drop tables and files into Microsoft products like MS-Word or MS-Excel (Microsoft 2006). STATA (version 7.0)[3] was used for most of the multivariate statistics because, on the one hand, a survey function is implemented by which the sampling procedure can be taken into account (this is also included in more recent versions of SPSS). On the other hand, STATA also has a broader set of statistical procedures implemented, which are of importance in the context discussed here. Whereas linear or logistic regression approaches can be run in both software packages, the Poisson and negative-binomial regression procedures are directly implemented in STATA, whereas SPSS needs an additional (costly) module. Other procedures like quantile regressions – not directly relevant in the context discussed here – are not implemented in SPSS at all. Both packages have their advantages and disadvantages. The final decision as to which type of software is preferable – next to the just mentioned restriction of the possible analyses – is left open to any researcher and may first of all depend on availability and preferences.

This chapter is structured as follows. Section 2 deals with a random sample of applicants of the priority year 2000. After the discussion of the sampling procedure and the data editing, some descriptive statistics are provided and a logistic regression approach is applied. Section 3 describes the matching of the R&D-Scoreboard (DTI) with the EPO database on patent filings. After a presentation of the problems and obstacles that were encountered during this process, some descriptive results of the analysis of this combined dataset are provided.

2 Random sample

The first approach to the question of linking economic and patent data presented here is a random sample of patenting firms at the European Patent Office (EPO), which was combined with some further information on the applicants that can be found in international company databases. This mi-

---

[2] See http://www.spss.com.
[3] See http://www.stata.com.

cro-level random sample of about 1,000 patenting firms is tested for its representation of the population of applicants at the EPO in the priority filing year 2000. Furthermore, the data are used to analyse the structure of the patentees and the dependencies of their patent activities on structural factors. Thus, it is not intended to give quantitative results of future filings at this stage of the work.

2.1 Databases and their characteristic features

2.1.1 The applicants' sample at the EPO

It is known from many analyses and studies that the distribution of applicants is highly skewed. Namely, the great majority of applicants applies for only a few patents each, whereas the top 150 applicants are responsible for the lion's share of applications. Blind et al. (2004) were able to prove, at least for German applicants, that the top 3.16% (first interval of the geometric mean) account for about 52% of all applications in the year 1999 and the top 10% account for about two thirds of the patents. They also showed that this skewness increased during the 1990s. Even though the number of applicants increased within this time period, the concentration increased too.

This skewness had to be taken in account for sampling as a pure random sample on the basis of all applicants would have been a sample first of all of the smaller applicants. Though they are the majority of applicants, they are not responsible for the majority of applications. Furthermore, it is a justifiable assumption that applicants' needs and conditions change with increasing size (in terms of patents) and each of these groups has to be represented in the sample in a sufficient number. So it is helpful to draw a stratified sample, which serves the purposes of the analysis much better and which takes into account the specialities of the population.

Several good reasons exist, why stratification is used in sampling procedures. Lee et al. (1989, p. 11) name the following four reasons:

"(1) the sampling variance can be reduced if the strata are internally homogeneous; (2) separate estimates with predetermined precision can be obtained for strata; (3) administration of the field work can be organised using the strata; and (4) different sampling needs can be accommodated in separate strata."

Whereas the second and third reasons are of minor importance for the purposes discussed here, the first and the fourth decide the issue at this point. Though a proportional stratification always leads to a reduced or equal variance compared to a similar simple random sample, this is not guaranteed for disproportional sampling (Kalton 1983, p.21; Lee et al.

1989, p. 15). Furthermore, in most cases it is not necessary to apply the stratification approach, as long as a proportional sample is drawn, as a simple increase in the number of observations may have the same effect. But in this case, where disproportional stratification is used, it is appropriate, because the applicants are distributed unequally over the total scale of applications per applicant. To say it in other words:

"If the differences among the factors ... are large, optimum allocation may yield large gains, much larger than proportional allocation. This can occur for characteristics which are distributed with great inequality in the population, often in highly skewed distributions, if good information is available for separation into strata." (Kish 1995, p. 94)

Let N be the total population size. When a stratification approach is applied, with H distinct strata containing a population size N_h in each stratum with stratum sample mean $\bar{y}_h$, the formula for the overall sample mean is

$$\bar{y}_s = \sum_{h=1}^{H} w_h \bar{y}_h \qquad (1)$$

The formula for the total sample variance is

$$Var(y) = \sum_{h=1}^{H} w_h^2 (1 - f_h) \frac{s_h^2}{n_h} \qquad (2)$$

where w_h is the proportion of stratum h in the population, with $w_h = N_h/N$ and $\Sigma w_h = 1$. ($1-f_h$) is the so called finite population correction (fpc), where f_h is the selection probability of the elements in stratum h, defined as n_h/N_h. When the population is very large and the sampling fraction is small, the fpc can be neglected. Equation (1) indicates that the overall sample mean is simply a weighted sum of the means of the strata. Equation (2) shows that the total variance in the sample is defined as a weighted sum of the variances within the strata, whereas the variance between strata has no impact on the total variance (Kish 1995, p. 88).

After the decision that the final sample should contain 1 000 applicants, these 1 000 units were distributed among the four groups of sampling strata, defined by the number of priority forming patent applications in the year 2000. The total number of observations did not in fact turn out to be 1 000 but was reduced to 992. This is because eight double counts in the original sample provided by the EPO were found, which were then aggregated to one case each. The sizes of the four strata were:

- 300 applicants with 1 patent
- 300 applicants with 2–3 patents
- 300 applicants with 4–30 patents and
- 100 applicants with more than 30 patents.

As these four strata do not reflect the shares of these groups in the population, it is a disproportional and not a proportional stratified sample. Within any stratum, the sampling units have been drawn randomly. The selection of the strata and the fact that they are more or less equally distributed within the sample, but not equally distributed within the population, has no negative effect on the outcome as long as the criteria of randomness of selection is met. The final sample then is the combination of four random samples, which again is a random sample.

"The strata may in fact be created in any subjective way without risk of bias in the survey estimators; the use of probability sampling within strata protects against selection bias." (Kalton 1983, p. 27)

For sampling purposes, the stratification was based on the number of filings in the base year 2000 forming four groups or strata. This procedure can be called pre-stratification. However, in a first step of the analysis it turned out that the country of origin is an important and decisive factor for which we have complete control and information. That is why this factor was introduced as a second layer of stratification. Hence, it was not directly applied in the sampling process, but ex post. So the final stratification procedure is a combination of pre- and post-stratification. For future applications, it is intended to use both layers in advance so that only pre-stratification is used. Though a post-stratification approach is a valid and legitimate procedure, pre-stratification gives more control over the sampling units and allows for sufficient observations even in small groups. For an overview on stratification procedures including post-stratification, refer – for example – to Kish (1995).

Due to the disproportional stratification of the sample, it is not "representative" for the whole population of applicants at the EPO or, to say it differently, it does not reflect the distribution of the total population. To get a "representative" sample and to make statements or "projections" for all applicants in the year 2000, weights have been calculated on the basis of the real number of applicants per size group <u>and</u> country.[4] This weight-

---

[4] The number of countries in the analysis was restricted to the following 11 countries and groups of countries to reduce the scattering: Germany; France; UK; Japan, US; Italy; other Central Europe; Eastern Europe; other Asian countries (also including Turkey); a group consisting of Canada, Australia, New Zealand and South Africa; South America.

ing procedure is a two-steps-in-one approach, as a "recalibration" towards the original distribution of the population is done and a projection to the whole population of the year 2000 applicants is included as well. A distinction of these two steps would have been possible, but is not necessary for the questions addressed here.

The selection probability of any unit in stratum h is defined as $f_h = n_h / N_h$ and the weights are simply the inverse of the selection probability $g_h = 1/f_h$ for each unit in the stratum h, which then gives a number of cases and a distribution of the sample similar to that in the population. The weights g_h for the sampling units should not be confused with the weights w_h for the strata, which are calculated as $w_h = N_h/N$. As this projection to the total population has a direct impact on standard errors and significance levels, it might be "downgraded" to the number of observations in the sample, so that the weights are calculated by

$$g_{s\cdot h} = g_h \frac{n}{N}.$$

As the interest of this study explicitly lies in the total number of patent filings in the population, the sampling weights are not downgraded, but this procedure should be considered for any analysis where standard errors or significance levels are analysed.

By applying these weights to the sample, a number of 46 065 applicants responsible for 143 728 patents in the year 2000 is reached, whereas the real number is 149 070, which is an underestimation of about 3.6 percent.

The weights for each country within each size group can be seen in Table 8.1. The weights are simply the number of applicants in the population (see "pop" column) divided by the number of applicants in the sample (see "sample" column), or – to say it more precisely – the reciprocal of their selection probability. As there are no applicants in the size group of 30 patents and more for the last three country groups, it was decided to increase the number of Japanese, US American and Italian applicants on behalf of these three applicant countries. The weights for these three countries were set as follows: 144/27 (Japan); 185/39 (US); 5/1 (Italy). This procedure ensures that the real total number of applicants is met by the weighting process.

Table 8.2 shows the estimated numbers of patents with the help of the weights – based on the pre- and the post-stratification – compared to the "real" number of patents for the selected countries and country groups. Except for Germany and France, the overall estimation of patent filings is pretty good for the larger countries. Further on, an underestimation for the rest of Central Europe is obvious. The reasons for these effects are not easy to find. One point may be that the German overrepresentation is accompanied by an overrepresentation of companies with a large number of patents. The underestimation of France and the rest of Central Europe seems to be based on the underrepresentation in the group of 2–3 patents. As already mentioned, the introduction of the country of origin in the sampling approach should increase the fit per country even more.

Table 8.1. Weights of applicants per country and size group

	1 patent		2–3 patents		4–30 patents		30+ patents	
	Sample	Pop.	Sample	Pop.	Sample	Pop.	Sample	Pop.
DE	52	4256	51	1301	47	755	14	79
FR	15	1750	16	469	17	250	7	28
GB	13	2184	11	516	19	248	1	12
JP	24	1844	19	642	30	557	27	135
US	89	9646	114	2846	122	1907	39	182
IT	13	1674	14	367	8	149	1	4
Other Central Europe	47	4808	27	1196	41	656	9	55
Eastern Europe	12	780	10	117	3	16	---	---
Other Asia + TR	12	2633	15	543	5	164	---	9
CA, AU, NZ, ZA	18	2257	16	407	3	161	---	3
South America	4	370	4	68	3	20	---	1
Total	299	32202	297	8472	298	4883	98	508

Source: EPO Sample.

Table 8.2. Estimated and "real" number of patents per country in the year 2000 for the total population of applicants at the EPO

	Estimated N of Patents	"Real" N of Patents	Absolute difference	Rel. difference ("real"=100%)
DE	28927	24590	4337	117.6
FR	6663	8212	-1549	81.1
GB	6271	6263	8	100.1
JP	23494	23548	-54	99.8
US	48237	50107	-1870	96.3
IT	4147	3927	220	105.6
Other Central Europe	14912	19358	-4446	77.0
Eastern Europe	1183	1113	70	106.3
Other Asia + TR	5238	6453	-1215	81.2
CA, AU, NZ, ZA	3975	4761	-786	83.5
South America	679	738	-59	92.0
Total	143726	149070	-5344	96.4

Source: EPO Sample.

2.1.2 Adding further company information

On the basis of the stratified sample stemming from the European Patent Office database, it is possible to analyse some structures and driving forces of patent filings. But there is no further information on the economic situation of each applicant in question, which may reveal further causalities on patent applications. Therefore, searches for further items in international firm databases have been conducted, namely sector of main activity, sales per year and number of employees. This information is rather common and therefore accessible fairly easily for large numbers of firms, whereas input indicators of the innovation process, like R&D expenditures or R&D personnel, are hard to find in publicly available data sources in a reliable manner.

Different sources were used in hierarchical order. These are: (1) Dun & Bradstreet's "Market Europe" and "World Base"[5], (2) Creditreform[6] data on companies from Germany, Austria and Switzerland; and Hoppenstedt[7] company database for Germany. For the companies still open, a classification at least of their sector codes was done manually with the help of internet searches.

The sector classification for some companies was rather unusual, especially for larger firms, which were assigned to the service sector (e.g. holding) instead of activity within the manufacturing sector. Siemens, for example was classified that way. This led to an improbable distribution of patents over sectors. This is why these improbable sector classifications were revised manually in a second loop. Siemens is now classified as a manufacturing company active in the field of electronics, for example. This is a class which surely does not reflect all the activities of this company but which points to the core of its products. As with any one-dimensional classification, assigning one sector only is always a reduction of information and a simplification. It will become clear in the further analysis that the sector variable has an important impact on the filing activities.

Furthermore, within these databases the different factors have different probabilities of being available, which means that there are different propensities for missing items. Whereas the sector information is unproblematic, the number of employees is less often to be found, and even this is more commonly found than information on the annual turnover. All these

---

[5] See for example: http://dbgermany.dnb.com/English/default.htm; or: http://www.dnb.com/us.
[6] See http://www.creditreform.de.
[7] See http://www.hoppenstedt.de.

activities – with all these mentioned problems– led to the following coverage concerning the three additional variables:

- sector: 923 companies (93.0%)
- employees: 672 companies (67.7%)
- sales: 558 companies (56.3%)

2.2 Descriptive statistics

Table 8.3 displays the distribution of applicants and applications by sector after the weights are applied. Only 10.4% of the applicants could not be identified at all in the company databases or via internet searches. Furthermore, this group only accounts for 4.6% of the applications. This reflects the fact that more smaller companies and free inventors could not be found as they are underrepresented in such company databases.

A concentration on the column with the valid percentages reveals that about 46% of the applicants can be assigned to the manufacturing sector, followed by 27.7% free inventors and 11.9% of public research and technical services. This is in line with the experience from other studies. Schmoch and Koschatzky (1996), for example, found shares of free inventors at the German Patent and Trademark Office in the 1980s and 1990s of more than 20%.

Looking at the total number of patents – euro-direct + euro-PCT –, the hierarchy is kept, but the shares are somewhat different. A clear majority of nearly 70% have their origins in the manufacturing sector, followed by public research and technical services, as well as by free inventors, who account for about 9% of the applications. These results indicate that the applicants from the manufacturing sector have – on average – a higher number of applications, whereas especially free inventors account for only a small share in terms of applications, compared to their share of applicants. Service companies apply for less patents than could be expected on the basis of their share of applicants. Their contribution at about 4.5% is rather small. This result is confirmed by a study for the European Commission, where the share of service patents lies at about 3% by a top-down approach of the largest applicants in this sector (Blind et al. 2003b).

Table 8.3. Applicants and applications in the year 2000 by sector

	Applicants			Applications		
	N	%	valid*	N	%	valid*
Agriculture, Building, Mining, Energy	547	1.2	1.3	1685	1.2	1.2
Manufacturing	19161	41.6	46.4	94343	65.6	68.8
Public Research, technical services	4898	10.6	11.9	14570	10.1	10.6
Transport, Communication	293	0.6	0.7	1601	1.1	1.2
Services	3738	8.1	9.1	6162	4.3	4.5
Holdings	582	1.3	1.4	3310	2.3	2.4
Public Admin.	660	1.4	1.6	2900	2.0	2.1
Free inventors	11375	24.7	27.6	12557	8.7	9.2
Total valid	41253	89.6	100.0	137128	95.4	100.0
Not classified	4812	10.4		6600	4.6	
Total	46065	100.0		143728	100.0	

* valid means: non-missing values.
Source: EPO Sample.

There seem to be differences between the sectors when looking at how they file their patents (not displayed here). For firms from the manufacturing sector and also for holdings, the euro-direct and the euro-PCT are nearly on the same level. Public research and technical services as well as the communication branch and public administration obviously prefer to make PCT-applications rather than direct filings at the EPO (or national first filings). One reason may be that companies in these sectors, outside the group of traditional patent clients of the production sector, serve international markets much more than the traditional ones. They are not just active in the European market or in some national markets within Europe, but they act globally and apply for patents also in other continents and global regions.

Table 8.4 contains the numbers of employees grouped in six classes. It is interesting to notice that nearly 65% of the applicants are small or medium-sized enterprises (SME, defined as companies with less than 250 employees), whereas they account for only 25.5% of the patents at the European Patent Office of the priority year 2000. It can also be seen that the skewness is kept when the number of employees is taken into account. The largest companies with more than 2 000 employees reach a share of 14.5% of applicants and 56% of applications.

The distribution of applicants and applications by sales is displayed in Table 8.5. Due to missing information – item non-response – in the databases, the share of companies without any information on their annual sales is relatively high. But, compared to response rates in "normal" surveys of 15–30% and additional item non-response, these values are still high, even though this may result in some bias towards larger companies. The EPO regularly reaches an overall response rate of about 35% in their recent Applicant Panel Surveys (EPO 2006).

Table 8.4. Applicants and applications by number of employees

Employees N	Applicants			Applications		
	N	%	valid*	N	%	valid*
1-9	4769	10.4	19.2	7843	5.5	7.0
10-49	4879	10.6	19.6	8776	6.1	7.9
50-249	6445	14.0	25.9	11871	8.3	10.6
250-499	2196	4.8	8.8	6037	4.2	5.4
500-1999	2958	6.4	11.9	14575	10.1	13.1
2000+	3603	7.8	14.5	62418	43.4	56.0
Total valid	24851	53.9	100.0	111520	77.6	100.0
Not classified	21214	46.1		32210	22.4	
Total	46065	100.0		143730	100.0	

* valid excludes missing values.
Source: EPO Sample.

Table 8.5. Applicants and applications by sales (grouped)

Sales EUR million	Applicants			Applications		
	N	%	valid*	N	%	valid*
1-199	1054	2.3	5.3	1790	1.2	1.8
200-999	3081	6.7	15.4	4477	3.1	4.5
1,000-9,999	4580	9.9	22.9	8366	5.8	8.4
10,000-49,999	4625	10.0	23.1	10173	7.1	10.2
50,000-99,999	1544	3.4	7.7	5111	3.6	5.1
100,000-999,999	3561	7.7	17.8	19279	13.4	19.3
1,000,000+	1582	3.4	7.9	50601	35.2	50.7
Total valid	20026	43.5	100.0	99797	69.4	100.0
Not classified	26039	56.5		43930	30.6	
Total	46065	100.0		143727	100.0	

* valid excludes missing values.
Source: EPO Sample.

About 44% have annual sales below 10 million euro and about 8% claim to have sales of one billion euro and more per year. Again, in terms of applications the shares are completely different. Companies with less than 10 million euro account for about 15% of the filings and the largest group applied for about 51% of the patents, given that the non-response is equally distributed among the sales classes.

2.3 Continuous vs. discontinuous applicants

One important step in forecasting future filings is the answer to the simple question: Will an applicant make future filings at all? In this section, a logistic regression approach (Aldrich, Nelson 1984; Liao 1994; Long 1997; Pampel 2000)[8] is used to estimate the probability of having an application in 2001, given the various items of structural information contained in the combined database of patents and economic data. The research question is: What is the companies' probability of applying for at least one patent in the year 2001 after controlling for country, number of employees and number of patents in the year 2000? One important thing has to be kept in mind: the stock of applicants, the population, is fluid and changes from year to year. But there are continuous and discontinuous applicants. Our sample of the year 2000 is not representative of any other priority year except for 2000. Thus, what can be done here is simply to estimate the probabilities for the members in our sample to be a continuous or discontinuous applicant, when looking one year in the future.

In Fig. 8.1, the results of the estimation of the logistic regression for four groups of countries (large European countries, Japan, US and others) are displayed. All models described here are so-called full-factorial or saturated models including all main effects and interaction effects. As the interest does not primarily lie in modelling with reduced degrees of freedom, but in detecting the overall influences on the outcomes, this approach is preferred to more simple models (in terms of degrees of freedom). The dependent dummy variable – having an application in 2001 (1) or not (0) – is explained with the help of the applicant's country and number of patents in the year 2000. As can be seen, the probabilities continually increase for all four groups, but with different slopes and starting from different levels. Whereas the companies from the large European countries (DE, FR, GB, IT) – of which it can be said that they have a home advantage at the EPO

---

[8] The sample design – disproportional stratification – was taken into account for these analyses. As only the standard errors and not the coefficients are affected, the discussion of the impact of this sample design is omitted.

compared to companies from non-European countries – start with the highest probability when having only one patent in 2000, their slope is less steep and the curve reaches perfect determination not before a number of 19 patents in 2000. This number tells us that companies from large European countries file at least one patent in the following year (2001) for sure, if they have applied for 19 patents in the previous year (2000). The increase of the probabilities – the slope of the curve – is different for the four regions of the world. The Japanese applicants show a very steep development of the curve, indicating that they are a continuous applicant for sure at a level of 9 patents per year. It seems that once these companies decided to serve the European market, they do this continuously, which is not a very surprising result as any market opening always comes along with enormous investments that follow a strategic management decision. However, this holds more or less for all countries – especially outside Europe. Applicants from small – often technologically specialised – economies strongly focus on international – here European – markets, as they do not have large national markets. One restriction (or extension) to the result presented here is that applicants from these countries first of all use euro-PCT applications much more often than euro-direct applications; and they use them much more frequently than companies from all the other countries. This indicates that they are not only directed towards the European market or some markets within Europe, but towards markets all over the world (or coming from markets all over the world). Thus, this last argument also holds for companies from the US and Japan, as they do not have a certain home advantage at the EPO or any apparent affinity towards Europe. Their lower probabilities to be a continuous applicant at the EPO must be guided by some other factors. And this is – first of all – the overall orientation towards foreign markets, as they have their own large national markets. From many studies it is well known that in particular the US American firms show a very selective activity in Europe (Fraunhofer-ISI et al. 2002; Grupp et al. 1996). They only apply for patents for technologies with a high international interest.

The slope of US-American applicants is the most moderate among the regions analysed here. This result is somehow extraordinary as in earlier years they showed a steeper curve. Reasons for this outcome can be found in a slightly changed behaviour after September 11th and mostly in the worldwide recession, especially driven by the tremendous downturn of the ICT-branches (information and communication technologies). Other countries that highly rely on information and communication technologies also showed a similar downturn in their patenting activities after the year 2000. A similar argumentation also holds for Japanese applicants. Within the 1990s they had to cope with the Asian crisis that uncoupled the Japanese development from the average growth at the EPO. In the late 1990s they caught up again and it even seems that they want to make up their backlog. Having this in mind, the interpretation of the total development of these countries has to be done with caution and with special attention when this data is used for forecasting purposes. The year 2000/2001 development may not reflect the overall trend properly.

The results of a second model that breaks down the probabilities of being a continuous applicant or not, based on sector information, are given in Fig. 8.2. Since differentiated sector information (going down to NACE/ISIC 2-digit level) is available, an aggregated version is used here, differentiating between five groups or sectors. This has been done in order to have enough observations to analyse the data in depth and also to make a distinction between the core and the periphery of patent applicants or – to put it in other words – between traditional patent clients in manufacturing and the others.

At the lower end of the number of applications in the year 2000, the companies from the manufacturing sector have the highest probability to be continuous applicants (Fig. 8.2). But the slope is not as steep as for the public research etc. sector, which reaches perfect determination at a level of 14 patents in the year 2000, whereas for the manufacturing sector this is at 22 patents. The energy/communications sectors has the flattest slope among the sectors under consideration here, reaching 100% probabilities of filing at least one patent in 2001 at a level of 40 patents in 2000.

Free inventors start at the lowest level and have a medium slope that sums up to a 100% probability at a level of 32 patents, but in fact there are no free inventors in our sample with more than six patents in 2000, so that any probability above this threshold is fictitious or virtual. This indicates that free inventors have a maximum chance of being a continuous applicant of about 25%. Firms from the service sector lie in between the other sectors.

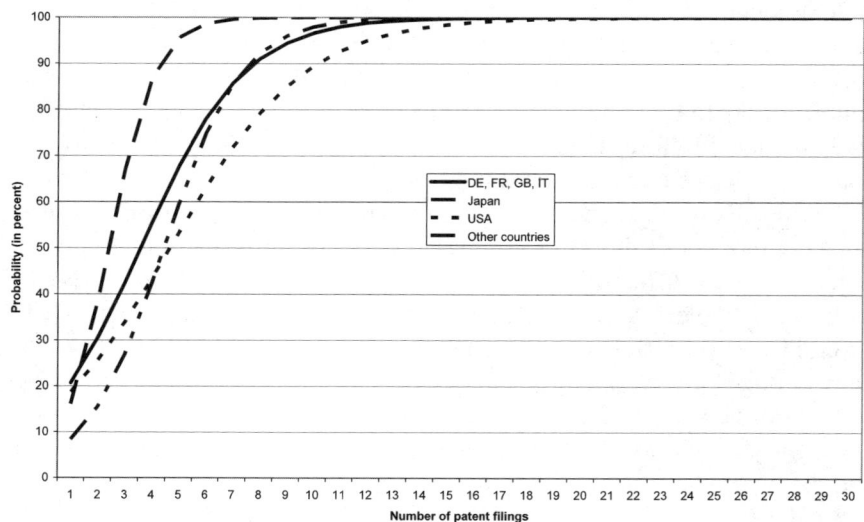

Fig. 8.1. Probabilities of having an application in 2001 (y axis) by country and number of filings in 2000 (x axis)

Source: EPO Sample.

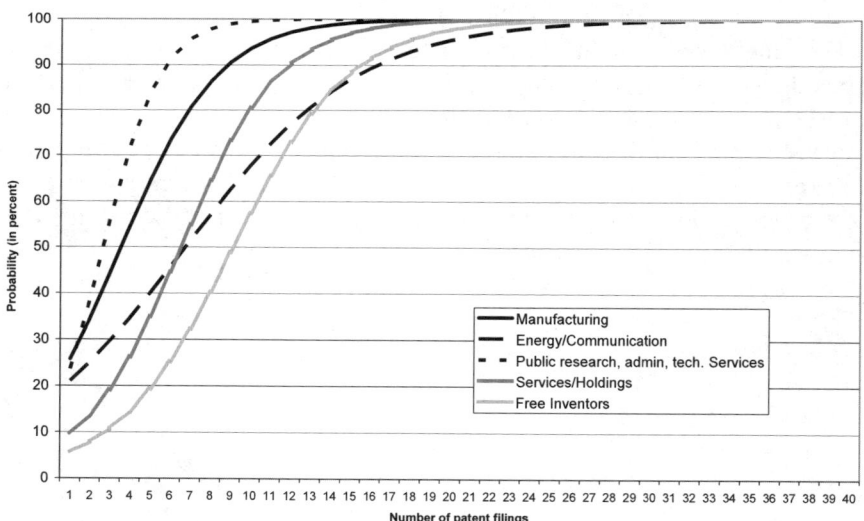

Fig. 8.2. Probabilities of having an application in 2001 (y axis) by sector and number of filings in 2000 (x axis)

Source: EPO Sample.

2.4 Discussion

In this section it was shown that it is possible to make substantial statements and come to meaningful results on the basis of a random sample of applicants at the European Patent Office – at least for one cohort. Thus the feasibility of a micro-approach has been proved.

The highly skewed distribution of applicants in terms of applications led to the approach of a stratified sample instead of a simple random sample. This stratified sample had to be weighted or redressed to the real distribution of applicants in the population. After doing this, it was possible to find plausible results and give some detailed descriptions of the applicant population. Besides, it was possible to show that there is a broad scattering over the different sectors ranging from traditional manufacturing firms – still responsible for the lion's share – to companies from the service sector.

In the second part of this section, the simple question of continuous and discontinuous applicants was assessed with the help of a logistic regression estimation. Strong indications were found that there are distinctions between them, which have to be taken into account both in sampling of populations as well as in forecasting procedures. The countries' activities seem to be guided – next to technological competencies – by further factors like having a strong home base and – backed by that – having a certain export portfolio or having a principal international orientation due to small national markets or strong technological specialisation. But this also leaves some space for the influence of overall economic developments that guide the behaviour of these applicants at least to some extent. Concerning sector differentiation, it was worked out that among the small applicants, the companies from public research and manufacturing sectors have the highest probabilities to be continuous applicants, while companies from the energy/communication sector have the lowest.

All in all, this kind of micro approach is very promising for the analysis of patent applicants. Furthermore, the results may help to better understand the driving forces of patent filings and give support for future forecasting activities of patent applications at the EPO. In a further first multivariate step, a linear regression model was fitted against the data and already shows some interesting results. But as there is over-dispersion (Long 1997) in the distribution, indicating that the factors are not normally distributed and that there is heterogeneity in the variance, a count model was fitted as the next step. Only preliminary results are available for this model and can not be reported here. For fitting these models, the survey design should be taken into account. Most statistical packages like SPSS or STATA are capable of these questions.

Though some crucial questions have been addressed, the approach and the analysis still have some limits, which are worth mentioning at this point and which leave space for further improvements. As only one cohort was analysed, a longer perspective or a repeat of the analysis could show the stability of the findings over time. Next to this, the inclusion of the differentiation between euro-direct and euro-PCT filings may lead to a further improvement of the results. A systematic analysis of different ways of application may reveal other distinctions, which then can be used in a forecasting process. A more detailed and disaggregated study of sectors and countries, which should accompany an in-depth comparison with the real distribution of the population, may offer some deeper understanding and some additional explanatory power.

For the application of this information in a forecasting exercise, three steps are necessary. Given a representative sample of a base year (2000 in this case), including data on patent filings of this sample in the following year, firstly, the shares and partial (by country, sector, size), probabilities of being a continuous or discontinuous applicant can be estimated. Furthermore, the number of filings of the discontinuous applicants can be calculated. Secondly, the number of filings of the continuous applicants has to be estimated in a regression approach, using a count model.[9] Finally, another random sample has to be drawn from the population of the following year (2001 in this case) and the just estimated coefficients can be applied to this sample, after the discontinuous applicants have been separated. To the estimated number of filings of these continuous applicants, the number of filings of the discontinuous applicants, calculated in step 1, has to be added.

There are two main assumptions underlying this approach. Firstly, the connection between the independent variables (country, sector, size etc.) and the dependent variable of patent filings is (more or less) stable over time. This implies that the coefficients estimated in the base year also hold for the sample in the following year. Secondly, the number of discontinuous applicants and the number of filings they account for, is (more or less) stable over time. Within this short time period of one year, these assumptions are not implausible.

The advantage of this approach, next to the usage of economic factors for the estimation of filings on the micro level, is that the inflowing and outflowing applicants are treated separately and their numbers of patents are not used in the projection process. There is no indication that inflowing

---

[9] But in the future, instead of computing two different models for each question, it might be appropriate to apply a zero-inflated negative binomial regression model, which can handle both of these questions in one approach.

applicants come in with a higher number of filings when they enter the process for the first time. There are some indications that the number of small and discontinuous applicants increased in the recent past, but this has only a limited effect on the overall number of filings when a short time period of one year is observed.

3 Combining the DTI Scoreboard with patent data

In a second approach to linking structural and patent micro-data, an existing database containing R&D-intensive companies – the so called DTI Scoreboard (DTI, 2001) is used and matched with the patent database at the EPO. This approach leads to a dataset which also contains R&D investment, number of employees, sales and profits. There are several problems in linking these two sets of data. In the first part of this section, the database and its construction, its specialities and the obstacles to combining the underlying source with the EPO database are described. In the second part some descriptive statistics of the combined database are presented.

3.1 Data

The UK Department of Trade and Industry (DTI) annually publishes information on the top international companies in terms of expenditures for Research & Development (R&D). The available information covers, besides sector, country and R&D expenditures, also figures for sales, number of employees, profit or market capitalisation. In the annual publication, R&D expenditures for the previous four year period are available too. Also some further data – in most cases in relation to R&D, sales or number of employees – is provided. In using more than one year of publications of the Scoreboard, it was possible to set up a database with some 650 companies and their activities back to 1991 starting from the basis year 2000, although not all information for all companies and all years was available, of course. However, there can be some standardisation problems when using these data. In the years before 2001, there were only 300 companies in the Scoreboard, in 2001 there were 500, in 2002 there were 600 and in 2003 the DTI increased the number of companies to 700.

The database contains the following main items:

- Name, country and sector.
- R&D expenditures (1991–2000).

- Sales (1994–2000).
- (Operating) profit (1994–2000).
- Number of employees (1999, 2000).
- Capital investment (CAPEX; 2000).
- Cost of funds (1994–1999).

When using the DTI Scoreboard, some problems were encountered, of which three are worth mentioning in this description. The first problem arises from the fact that data for different companies for several years is used. Within this time period – first of all because the largest companies in the world are addressed – some mergers and acquisitions took place. The question for the merged companies then is, how to treat them for the time before the merger? It was decided to sum up all the data for these firms and to treat them as merged from the beginning of the observation period. The reason is comparability over time, as no separation is possible after the merger. But this procedure involves some new problems, as the outcome of a merger is in many cases a reduction in costs and the realisation of returns to scale. So a reduction of R&D investment does not directly mean that their efforts in R&D are lower than before, but rather that the whole is not the sum of its parts. One further problem is that only information on mergers among the top 500 (300) companies in the world is available. If one of these companies merges, for example, with the number 501 on the (fictitious) list, no information about the sum of the parts before the merger is in the database. So an increase in R&D expenditures may, next to further efforts in R&D, also simply be due to a merger with a smaller company, which also did some R&D. Thus, concerning the research question of factors guiding patent filings, this is of lower importance as the number of patent filings for the entities in the year of merging is available in the database. The only problem arises when time lags are taken into account for the estimation of influences. In that case, such companies have to be excluded from the analysis. In the other cases, all the companies involved in a merger (where information on the merger exists) are treated as one company from the beginning, as already mentioned above.

One further problem exists, which is of certain importance for the linkage with the patent data. The companies in the DTI Scoreboard are in some cases just holding companies or groups, which do not apply for patents themselves, but their subsidiaries or affiliates do. This means that it is important to have – at least to some extent – a certain knowledge of the group structures of these companies. Any person who tried to figure out all the relations of large firms knows that this is a nearly impossible mission. A feasible solution had to be found for the purposes discussed here. Therefore, only the "obvious" or well-known relations are included, which in some cases have been paralleled by internet searches for such companies, where the structure is not that well known or obvious. Furthermore, it is not possible to identify all patent applications in the in-house database of the EPO for large companies or groups. On the basis of applicant numbers, the test led to an underestimation of the number of filings of some firms, for a number of different reasons. On the one hand, large companies may have research facilities in different countries, which then are treated as different applicants, unless the group or mother applies for patents. On the other hand, the EPO database does not contain any information on mergers and affiliates or subsidiaries. So, also from this perspective, the relations of the companies cannot be surely identified.

Solutions to all these problems could not be found within this project. A simple search by company names – for example – may help to overcome the problem of different applicant numbers for the same applicant, but the other obstacles still exist. As there is work in progress within the EPO to set up a database with information on the interweaving of the applicants, the quality of the data source may increase in the near future. In order not to waste too much effort on "finding the last patent" of each applicant, a mixed search strategy on the basis of applicant numbers and also name searches has been applied, backed by some internet queries for several firms.

One more factor has to be taken into account when matching external databases with the internal data at the EPO. The DTI Scoreboard assigns each company to one country – usually the country of the headquarters. On the other hand, again, only the largest companies in the world are addressed, which are – not by definition, but as a matter of fact – internationally active and globalised firms. A restriction to one single country is not appropriate. This is another reason why an enrichment of the query of applicant numbers by the search of applicant names was necessary, including applications originating from different countries of residence, if needed. Even so, in some cases still no patent application was found at all within the time period under observation. This is due to several reasons: one may be that the companies are active in sectors where patents are less important. Instead, another protection instrument may be used. Another reason may be that the firm has a certain strategy or export portfolio and does not apply for patents at the EPO but uses national offices to get protection in selected markets only.

Concerning the time perspective of the data, one more problem had to be faced. In the years before 2001, the "hit list" of the R&D-intensive companies contained only 300 firms. The number was increased in the year 2001 to 500 companies. With the help of a simple subtraction, it follows that data are not available for all 500 companies for all the years. Furthermore, over the years there has been "relegation" and "promotion" in and out of the premier league of R&D-intensive companies. This fact also leads to incomplete or lagged time series. Anyway, this has to be accepted, as the only solution would have been to collect all the missing data by hand or to apply a multidimensional imputation model. For the statistical analysis, our approach leads to changing numbers of observations, when looking at different years or when different time lags are taken into account, for example, for the effects of R&D on patent filings.

This approach, therefore, is far from being perfect, even if it is possible to start with good data. Three obstacles for the linkage of structural and patent data have been identified for this sample. Some consensus or practical solutions had to be accepted: first, the fact of mergers in the time period under observation has been solved by treating merged companies as one company from the beginning. This means that one point in time is fixed and the situation at this point is given, namely the situation in the year 2000. Secondly, not all company structures and all the interweaving can be taken into account. This holds for economic as well as patent data. Thirdly, complete information for all points in time are not available for all companies in the sample.

All the obstacles and restrictions are qualified, when a simple cross-sectional analysis is conducted, which is the original intention of this part of the study.

3.2 Descriptive statistics

In Table 8.6, the countries of origin of the 630 companies in the sample are displayed. It can be seen that firms from the US stand for about 38% of all observations, followed by Japanese firms, accounting for about 22%. 67 firms – or 11% – stem from UK, 7% from France and only 6% from Germany. This distribution was clearly different in Sect. 2, where the results of the random sample were discussed. The share of German applicants there was much higher – higher in fact than the shares of British and French applicants.

Table 8.6 also summarises the mean values of some selected variables for the year 2000, differentiated by the country of origin of the different companies. The result of the low representation of German applicants is qualified by the other figures of this table. The German companies are the largest among the firms examined here in terms of sales, number of employees, R&D expenditures and patent applications. Only the mean operating profit is higher in the other European countries compared to the German firms. The smallest firms originate in Scandinavia in terms of sales and number of employees or are located in non-European countries when R&D-expenditures and patents are taken into account, respectively. This is an effect, on the one hand, of the concentration of the companies within countries, emphasising the – on average – higher concentration and size of firms in Germany. On the other hand, this is also an effect of the national technological specialisation – the sector distribution of the firms from different countries.

Table 8.6. Means for selected variables of the year 2000 by country

	N	%	Means				
	Companies		Sales in £m	Operating Profit in £m	Number of employees	R&D-expend. in £ 000	EPO patent applications
FR	45	7.1	11019	1062	80214	369355	93
DE	35	5.6	13631	1533	94194	548012	225
JP	137	21.7	8307	346	40299	352275	91
GB	67	10.6	12420	1311	50924	322365	96
US	240	38.1	9109	1260	43325	435925	77
CH	21	3.4	7141	1314	51344	409999	88
Scandinavia	29	4.6	5377	896	33463	321287	166
Other European Countries	30	4.8	12010	1880	68979	338680	142
Non-European Countries	26	4.1	6553	1051	38418	238460	23

Source: DTI Scoreboards, EPO.

This last aspect can be seen in Table 8.7, where selected variables for the year 2000 are provided for the eleven sectors used here. An aggregated version of the original sector classification provided by the DTI is used, forming 11 groups out of the former 27 groups, done for statistical purposes and to obtain clarity. Among the world wide most R&D-intensive companies, 100 (15.9%) are involved in IT hardware production and distribution and another 11.6% are working in the branch of software, IT services, telecommunication and support. Nearly the same amount as with IT hardware are dealing with pharmaceuticals, health or personal care, followed by 10% which are involved in electricity, gas, oil or steel. The electronic and media branch is represented by 57 (9%) firms and another 52 (8.3%) observations stand for the chemical industry. The other sectors reach shares between 7.1% and 3.3%, with aerospace and defence at the lower end of the scale. This distribution also makes clear – as the countries' shares already suggested – that this database does not have the same structure as the random sample or the whole population of applicants at the EPO.

The smallest sector in terms of patents is – also in line with the expectations – software, IT services and telecommunication. This is less due to their limited effort in R&D as they reach a good medium position in this dimension. The reason is more easily explained by the fact that the EPO still does not accept software patents in the same way as – for example – the USPTO does. Furthermore, the companies from these sectors are much more active in services and a large share of products, goods and services is less often patentable at all. Besides this, they make much more frequent use of other protection instruments and intellectual property rights like trademarks and copyrights, of course.

In total, the 630 companies in the database account for 55 660 patents in the year 2000 at the EPO and 36 697 in the priority year 1995. This is more than one third of all applications in 2000 (and also in the previous years). Though one can not be sure that all applicants and all applications of the entities in the database have been identified – as was already discussed in the previous part of this section – this is a pretty large share for less than 1.5% of all applicants in that year. This figure is further qualified when it is taken into account that at least one patent in the year 2000 was found for only 497 firms, as Table 8.8 suggests. In fact slightly more than 1% of all applicants account for about 38% of all applications of the priority year 2000. These applicants are covered by this database.

Table 8.7. Means for selected variables of the year 2000 by sector

	N	%	Means				
	Companies		Sales Mio £	Profit in £m	Number of employees	R&D-expend. in £ 000	Total patent applications
Aerospace & Defence	21	3.3	10070	718	78332	434159	83
Automobiles & Parts	42	6.7	23739	1398	122077	943851	107
Beverages, Food, Forestry, Household etc.	44	7.0	8347	1029	52400	171558	59
Chemicals	52	8.3	5277	458	27019	215602	106
Construction, Diversified, Financial, Retailers	36	5.7	14223	1884	82837	184175	69
Electricity, Gas, Oil, Steel	63	10.0	21957	2918	48901	150527	31
Electronic, Media	57	9.0	7489	718	61200	430897	228
Engineering & Machinery	45	7.1	5092	323	37331	143301	40
Pharma, Health, Personal care	97	15.4	5052	947	27903	469453	92
IT hardware	100	15.9	7419	774	39468	601567	158
Software & IT services, Telecom, Support	73	11.6	5948	974	33967	249229	22

Source: DTI Scoreboards, EPO.

Table 8.8. Number and share of companies of the DTI scoreboard with and without patent applications in the priority year 2000 at the EPO

	N	%
No patents found	133	21.1
At least one patent	497	78.9
Total	630	100.0

Source: DTI-Scoreboards, EPO.

The R&D expenditures for the sample are characterised in Table 8.9. The number of firms for which information on R&D expenditure is available decreases with the time going back in the past. This is due to the fact that, in former years, the DTI included only the top 300 companies in its "hit-list" and the historical data provided in the Scoreboard of 2001, which was used as the basis, only reaches backward to the year 1996. This is the reason why the number of observations dramatically shrinks in 1995.

In the year 2000, each company in the sample of R&D-intensive firms has spent about £392 m for Research and Development on average, ranging from a minimum value of some £43 m to a maximum of £4.5 Billion. Whereas the maximum value decreases over time, the mean constantly increased since 1996. The mean in the years before is much higher since only the top 300 R&D-intensive companies are considered there. The mean is not comparable over time. When looking at the development of the maximum value over time, an increase between 1991 and 1996 and a decrease between 1996 and 2000, indicates that the largest companies reduced their effort in the second half of the 1990s, whereas the smaller ones among the top 500 increased their activities, proved by the increasing mean value.

Table 8.9. R&D-expenditures 1991-2000

Year	N	Min.	Max.	Mean	Std.Dev.
2000	493	42898	4552149	391898	677596
1999	477	13969	4752979	362052	626670
1998	472	9946	5288526	342182	602100
1997	454	3160	5489356	316393	574416
1996	432	2855	5957960	293265	563498
1995	278	114	5615142	379801	599413
1994	282	600	4710001	379559	567878
1993	300	300	4036619	350825	545276
1992	294	4000	3960969	350647	548028
1991	182	19800	3941220	492666	629736

Source: DTI-Scoreboards, EPO.

3.3 Correlation

In Table 8.10, the correlations between the numbers of patent applications and the R&D expenditures are displayed for the six years under observation here. A correlation analysis does not need an assumption of a clear causality to compute the values. Although there are studies which showed that patent applications have a positive effect on subsequent R&D expenditures (Geroski et al. 2002), the interest here lies in the influence of R&D-expenditures on subsequent filings. In Table 8.10, some indication for the first causality can be found, the correlations between patents and previous R&D expenditures are positive and highly significant. As can be seen, the correlation is highest for R&D expenditures and patent applications of the same year and decreases with increasing time lag. This holds for all five years of R&D data displayed here and continues in the earlier years not displayed in this table. This suggests that the most effective period of R&D expenditures on patent filings is the same year in the case of the largest companies worldwide.

This analysis neglects the fact that the connection between R&D and patent applications has become more loose in the course of the 1990s in many countries, when time series are used, which has been shown in several studies in the recent past (Janz et al. 2001; Kortum, Lerner 1997; Kortum, Lerner 1999). Blind et al. (2003a; Blind et al. 2004), for example, have been able to prove, at least for Germany, that strategic motives of applying for patents have become more important – especially for larger companies – in the second half of the 1990s. This means that the number of patent filings increased even more than the R&D expenditure did. Whereas these studies are looking at developments over time, this analysis focuses on different companies at several points in time. This may explain the increase of the correlation between R&D and patent filings between 1995 and 2000, as the ratio of the number of patents per £10 m of R&D expenditure increased in the sample from 1.9 in 1995 to 5.3 in the year 2000. This means that, in 1995, the companies in the sample received 1.9 patent applications per £10 m, whereas in 2000 they filed 5.27 patents – on average – if the same amount of money was spent. So the increase of the correlation coefficient over time reflects an increase in the homogeneity of the sample in how they file patents (also strategically), but not how much they file per R&D spent.

Table 8.11 contains the correlation coefficients for the total patent applications per year with the number of employees in 1999 and 2000, as well as with the market capitalisation in the year 2000. For these three variables, no further information about the development over time is available, as the DTI did not include these factors in previous years. Anyway, as can be derived very easily from the correlation matrix, the market capitalisation, though having a positive and significant effect, makes only a limited contribution to the explanation of the number of patent filings. The number of employees, however, shows a high correlation with the number of patent filings, though it is interesting to see that the coefficients decrease with every priority year starting in 1995. This is due to the fact that the patent filings dramatically increased in the second half of the 1990s, whereas the number of employees did not increase or was even shrinking, especially in the manufacturing sector, which still has the highest contribution to the overall patent output. But the connection between patents and employees still exists to a large extent. Though the number of employees is a simple index for the size of the company and therefore has an obvious impact on the number of filings, it cannot be ignored in any analysis. Furthermore, these results support the results of the former findings on the basis of the random sample (not reported here), where even higher coefficients have been found, as in that case the variety of possible values of the employee-variable was much larger (not only huge but the whole range of companies were analysed) and therefore the covariation with the number of patents was much higher, too.

Table 8.10. Correlations of patent applications and R&D-expenditures 1995-2000

Year of patent application	R&D-expenditures 2000	1999	1998	1997	1996	1995
1995	0.484(**)	0.476(**)	0.478(**)	0.474(**)	0.448(**)	0.411(**)
1996	0.538(**)	0.531(**)	0.523(**)	0.508(**)	0.460(**)	0.437(**)
1997	0.537(**)	0.529(**)	0.515(**)	0.490(**)	0.439(**)	0.412(**)
1998	0.540(**)	0.533(**)	0.510(**)	0.476(**)	0.427(**)	0.399(**)
1999	0.538(**)	0.530(**)	0.503(**)	0.470(**)	0.412(**)	0.392(**)
2000	0.528(**)	0.507(**)	0.484(**)	0.462(**)	0.408(**)	0.382(**)

** Correlation is significant at the 0.01 level.
Source: DTI Scoreboards, EPO.

Table 8.11. Correlations of patent applications with the number of employees and market capitalisation

N of patents in ...	N of employees 2000	N of employees 1999	Market Capitalisation in £m
1995	0.449(**)	0.401(**)	0.177(**)
1996	0.449(**)	0.413(**)	0.170(**)
1997	0.438(**)	0.401(**)	0.155(**)
1998	0.426(**)	0.397(**)	0.177(**)
1999	0.417(**)	0.385(**)	0.187(**)
2000	0.423(**)	0.383(**)	0.187(**)

** Correlation is significant at the 0.01 level.
Source: DTI Scoreboards, EPO.

3.4 Discussion

This section was concerned with the top companies in terms of expenditures for Research and Development. For this purpose, two databases were combined. On the one hand, the DTI Scoreboard, containing many economic figures and indicators for these companies and on the other hand, this database was matched with the patent database of the EPO. Though there were several problems incurred in doing this and several compromises had to be accepted, it was possible to face this challenge. But even for this limited set of (well known) world wide active companies, not all patents – for all subsidiaries and affiliates – could be found for sure. This raises some doubts on the overall coverage of the newly built database in terms of patent applications at the EPO.

The correlation statistics showed that the dependency of patent applications on R&D expenditures is obvious. Some evidence was found for the thesis that patent applications lead to subsequent R&D investment. Concerning the influence of R&D on patent filings, it seems that there is no time lag, which has to be taken into account. R&D and patent filings of the same (application) year correlate highest. Also the number of employees is highly correlated with the number of filings – a result which was already found in the previous section dealing with the random sample of all EPO applicants. Of course, these results are to a large extent simply a matter of size. Anyway, other variables – including a size dimension – did not correlate with the patent filings in the same way. Whereas sales reach a medium level of correlation, the contribution of profit or market capitalisation is much more restricted.

Section 3 differs in an important way from Sect. 2. Whereas Sect. 2 aimed at making statements that are representative for all applicants of the year 2000, Sect. 3 intentionally focused on a very small subset of applicants, namely the largest. It could be shown that these top 1.5 percent of applicants account for more than 38% of all patents of the priority year 2000. If just one of these companies changes its efforts, activities, portfolio or just its habits, the total development of patent applications – especially in some technological fields – may be influenced massively. This will not be the case – or at least to the same extent – if some smaller applicants change their habits. Furthermore, to be able to change or guide their activities, a critical mass is needed, which can be treated strategically. From this perspective, it is appropriate to focus one's efforts mainly on the largest applicants at the EPO (or at any patent office), not only because they account for a very large share of all applications, but also because in many respects they act as trendsetters for the smaller companies. The analyses and examinations conducted here are one way to bring them into the spotlight and to observe them with due attention.

A possible extension to the analyses might be to estimate the probabilities that the companies will increase or decrease their numbers of applications in the future. This can be done — similarly to the approach of Sect. 2.3 — with the help of logistic regression. The dependent variable is then not the absolute number of filings, but a simple dummy variable, becoming zero for shrinking and one for expanding application figures. This approach may help to overcome the problems of estimating the absolute values or changes. Besides this, the probabilities of being a constant or irregular applicant and the driving forces behind changing figures may be uncovered. However, in Sect. 2 a distinction was made between continuous and discontinuous applicants. As almost all companies at the top of the list in terms of R&D are continuous applicants, this distinction makes less sense here. With constant or irregular applicants the direction of potential changes in the number of filings can be assessed.

Additional knowledge and realisation can be expected from the distinction between euro-direct and euro-PCT applications. The previously mentioned strategic treatment of patents may also express itself in changing usage of these different ways of applying for patents.

What is missing here again is, of course, the implementation of this data in a forecasting exercise. Next to the just mentioned binary regression of increasing vs. decreasing/stagnating applicants, the absolute number of future filings can be estimated in a regression approach. First tests have shown that a linear regression approach may not be appropriate, even for this rather homogenous sample. Instead, a count model applying a negative binomial regression estimation seems more appropriate (Hausman et al. 1984). The coefficients of such a model can be applied to future samples of the DTI-Scoreboard, then leading to short-term forecasts of future filings. But for this purpose, next to an improvement of the patent identification method, a kind of "strategic coefficient", reflecting the changed and still changing meaning of patents as a strategic instrument has to be developed. As many other studies have been able to prove and as also found in this study, the connection between patents and R&D becomes looser, for whatever reason. With this fact in mind, it is even more important to look for alternative (economic) factors and variables that explain the number of patent filings, in addition to R&D.

4 Conclusions

All the other chapters in this book approached the question of forecasting future patent filings at the EPO on the macro level of countries or on the meso level of sectors or technologies. This chapter aimed at making use of micro data of applicants/firms as a source for future forecasting activities. Even though a direct application of the methods presented here to forecast filings was not possible, the outlines have been drawn and the foundations are laid to fertilise it in future exercises.

Arguments have been given as to why standard databases or ready-to-use data sources are not appropriate for the questions addressed here. Instead, it was necessary to build up extra databases. This was done by two approaches: first, a sample of 1 000 applicants of the priority year 2000 and second, the analysis of the worldwide largest companies in terms of R&D expenditure. Each approach has its specific advantages as well as disadvantages.

It was shown that a sample of applicants – based on an adequate sampling procedure – is a feasible method to reach a source of information that allows progress to be made beyond standard results. It was one of the central arguments of Sect. 2 that it is necessary to understand the driving forces behind the application process, before a forecasting exercise is of any value to answer the question of the numbers of future filings. For example, the companies from smaller European and non-European countries have a higher affinity towards international markets, as they do not have large domestic markets. Thus they have no home advantage, even if they have a strong home base. This fact makes them even more susceptible to international crises and economic downturns. The patent activities at the EPO of companies from US and Japan are much more guided by their export structure and portfolio, as they have large national markets.

Furthermore, the continuity or discontinuity of the applicants' activities at the EPO have been addressed. For example, while applicants from the manufacturing sector have a high probability to file continuously, public research institutions have an even higher probability. Besides, the recession and the crisis of the ICT-sector also had an influence of the probabilities for filing continuously, first of all of the energy/communication sector but also of the manufacturing sector.

These results drawn from the random sample of applicants hold for the all applicants "on average". As the numbers of filings per applicant is highly skewed, it might also be appropriate to look at the largest applicants only. This was done in Sect. 3, where some additional knowledge on the behaviour of the group of R&D-intensive companies could be generated.

The analysis of this data showed, for example, the well-known result of a strong dependency of patent applications on R&D expenditures. Besides, some evidence was found for the thesis that patent applications lead to subsequent R&D investment and, concerning the influence of R&D on patent filings, it seems that there is no time lag which has to be taken into account.

However, concerning both approaches, future studies will have to establish whether or not clear and stable patterns of relations can be found. This is, of course, a crucial precondition for any application of these approaches in forecasting exercises. Both approaches promise high value for short-term estimations of patent applications, whereas mid- and long-term forecasts may not be the specific strength of these methods. The reason for this is very simple and forms the biggest advantage of the micro approaches: Much more information and variance can be taken into account, which is not visible at the meso or macro level. However, to file or not to file, that is the question that always has to be answered on the micro level of individual applicants.

Chapter 9

Improving forecasting methods at the European Patent Office

Peter Hingley and Marc Nicolas

Controlling Office, European Patent Office, Munich, Germany

Software used: Excel, SAS, Scientific Notebook, S-PLUS

1 Introduction

In the previous chapters, a spectrum of techniques has been developed for improving the routine forecasting methods for patent filings at the EPO. After discussing the existing approaches (Sect. 2), we will assess the aspects of each research contribution that seem useful for our purposes (Sect. 3) and will then indicate some results that we have obtained since the projects were completed (Sect. 4). The comparison of results from different forecasting methods is discussed in Sect. 5. Some possible alternative forecasting approaches are mentioned in Sect. 6, and Sect. 7 concludes.

2 Existing approaches

The EPO has previously used several methods in its effort to forecast the numbers of patent applications. Technical details of estimations made in early 2003 are given by Hingley and Nicolas (2004). Trend analysis is carried out mainly by using linear models and some simple time series methods. Regression methods are also used to construct a transfer model that takes into account the existence of first filings in the country of origin up to 12 months before filing at the EPO, under the terms of the Paris Convention. There is also a regular survey of applicants to determine future filing intentions. Pragmatic approaches are used to provide forecasts at the level of EPO joint clusters.

2.1 Trend analysis

Forecasts are made by extrapolating observed trends. Straight line regressions are taken over time for the historical yearly series. Systematic tests are carried out on series of various lengths, using linear models that incorporate linear, quadratic, and third order polynomial terms as well as simple nonlinear logistic (S shaped) models. First order AR(1) methods have also been used, since these can be relevant even when the underlying process is more complicated (Hendry 1995, p. 204).

The selection criterion for the competing trend based models is the goodness of fit of each model to the data. Formulae exist for deriving confidence intervals for out-of-sample forecasts that are made using linear models. These can also be easily adapted to logistic models (e.g. Draper and Smith 1981, p. 30). The confidence limits typically have the property that they widen to some extent as the forecasts move further forward in time.

One problem with this approach is that models can not easily deliver indications of future disruption of a recent trend. Some insurance against this is provided by fitting models back over many years of the available data, but this involves the assumption of an equilibrium trend that the process reverts towards after short period jumps. On the other hand, when models are fitted over recent history only, as in the ECM approach that was discussed in Chap. 3, there is more responsiveness to current movements and the problem of under-forecasting an explosive growth phase is reduced.

Various assumptions can be made, both about the structures of models to be used and about the lengths of historical data on which to fit the models. When taken together, these generate a "cloud" of trend projection lines. In practice, the process of finding a consensus consists in plotting a line through the densest area of this cloud. If formal confidence limits are calculated for any one projection method, these can be rather wide. They are possibly also incorrect, because the detailed error structure of the data has not been taken properly into account in the formulation of the model.

2.2 Transfer model

International patent law is based on the Paris Convention of 1883 (WIPO 2006), which specifies that a subsequent filing for an invention can take place within one year of the first filing (See Fig. 2.1 in Chap. 2). In the transfer approach, the numbers of EPO patent applications in a particular year are forecasted by using observed levels of world wide first filings that were made in the previous year. However, where the numbers of such first filings are not reported by a national patent office, the numbers of domestic national filings (excluding PCT national phase) can be taken as a proxy for first filings.

About 80% of world wide first filings are made in the three blocs: EPC contracting states, Japan and US (EPO, JPO, USPTO 2005). Attention is restricted initially to activities in the three main blocs: EPC, Japan and US. Annualised numbers of first filings (FF_j) in each of the three blocs (j = 1, 2, 3) are reported and extrapolated by trend analysis, or by obtaining forecasts from the various patent offices.

The corresponding numbers of EPO subsequent filings (SF_j) that originate from each of the blocs are obtained. From each bloc j, for each year t, transfer ratios (TR_{jt}) are calculated as:

$$TR_{jt} = \frac{\text{EPO subsequent filings in year t from bloc j } (SF_{jt})}{\text{First filings in year (t-1) in bloc j } (FF_{j(t-1)})}$$

The TR_{jt} are then themselves extrapolated by trend analysis over the planning period. From these projections, projected levels of SF_{jt} are calculated by reversal of the above equation, $SF_{jt} = FF_{j(t-1)} \times TR_{jt}$.

It is possible for clients to make their first filings at the EPO as well as in the national systems. This mainly occurs for EPC based applicants, for whom the EPO is an alternative to a national office for a FF. The numbers of first filings at EPO (FF_e) are projected, again using self determining straight line regression. Total filings at the EPO (first + subsequent combined) from other countries outside the three main blocs (TF_o) are also projected by linear regression. Finally, it is necessary to add the projections for TF_o and FF_e to the predicted SF_y values from each bloc in order to get forecasts for total EPO filings (TF_y).

$$TF_y = TF_o + FF_e + \sum_{j=1}^{3} SF_{jt}$$

This is a pragmatic first order method to model patenting at the EPO by exploiting the predictive power of the FF series. An earlier version of the transfer method is described by Hingley and Nicolas (2004), in which the sum of first filings in the three blocs EPC, Japan and US is used to build a single transfer ratio, rather than making an analysis of transfer rates from each separate bloc. These days both methods are carried out to see which performs best. Modified versions of the methods can be applied at other patent offices to forecast their total filings from world wide first filings.

In the future, it is hoped to develop the method by using more sophisticated methods of extrapolation (time series methods and multivariate analysis) and by taking account of further subdivisions of the data (such as the distinction between PCT and non-PCT filings, since PCT filings have a higher propensity for international transmission). Globalised versions of such filing flows models can also be considered (see Sect. 6).

However, there are some problems in interpreting patent filings data from other patent offices for use in the transfer model. Generally, the transfer ratios between offices have increased since the mid 1990s (see Table 9.2 in Sect. 6). But the first filings data at USPTO are subject to several factors that make them difficult to count. The US has used a first-to-invent rule over the period of the historical data, which means that the records of the invention itself can effectively establish the date of a priority right rather than the date of the first filing itself. There is also a provisional filings system in US that, to some extent at least, obscures the first filings counts. This may explain the marked increases in the apparent transfer rates in Table 9.2 from US to EPO and from US to JPO between 1997 and 2001.

For economists, it may be more interesting to consider transfer models based on the blocs (or countries) of residence of inventors rather than the blocs of residence of applicants. However, there are considerably more missing historical data on inventors than on applicants, which is the main reason that we prefer an approach that is based on applicant countries.

2.3 Surveys of applicants

Since 1996, the EPO has carried out an annual survey among applicants to profit from information about their intentions towards filing patent applications. The study is called the *applicant panel survey*, but this does not represent a panel in the econometric sense and is also not strictly a panel in the market research sense, since sample members change to some extent from year to year. For methodology and results of the annual surveys from 2001 to 2005, see EPO 2006. Since 2001, the survey has been carried out with the assistance of Roland Berger Market Research.

The selected applicants are asked to complete a questionnaire to give their intentions for filings in all major world patent systems for the previous year, the current year and for the following two years. Breakdowns of filings at each office are asked for in terms of first and subsequent filings. For the EPO species, a further breakdown is requested in terms of euro-direct and euro-PCT International Phase (euro-PCT-IP filings). An option is provided for the respondent to give qualitative predictions (more/same/

less), but most respondents in fact provide the requested quantitative information.

Since 2001, the survey has been carried out on an increased sample, that is based on two groups:

1. Biggest group. The largest 400 applicants at the EPO, in terms of the numbers of filings made during the previous year.
2. Random group. A group of about 2 000 applicants that is obtained by simple random sampling of European patent applications, where the probability that an applicant is selected for the sample is proportional to the number of applications made during the previous year.

The main purpose of the survey is to estimate a growth index of intended numbers of applications for the forecast horizon two years ahead. Different methods are used to estimate the growth indices for the two groups, and it is important to establish whether the projected behaviour of the biggest applicants is reflected in the overall population that is represented by the random group. Since the random group has been selected by statistical criteria, it is possible to use the results from this group to ascribe confidence limits to the resulting forecasts for overall filings by the population of applicants.

The existence of the biggest group, and the sampling method used for the random group, ensures that most of the largest applicant companies are surveyed. Many of them also make applications to patent offices other than the EPO. Information is also elicited on other items that can assist an understanding of the patenting process by companies (e.g. R&D expenses and numbers of patentable inventions).

Parallel surveys are carried out from time to time by JPO and USPTO. The comparison of results from the surveys of different offices provides an important potential control check. The EPO survey is an important forecasting method and is discussed further below, particularly in Sect. 3.7 where it is compared to another microeconomic approach.

2.4 Planning at the level of joint clusters

In the research programme, it was not intended to study separate technical areas in depth. Nevertheless, this is something that is now considered in the routine EPO forecasting procedures. Results from the applicant panel survey are used and trend analyses are made at the level of the 14 individual joint clusters, to determine plans for workload requirements as well as for future patent filings[1]. Bottom-up forecasting of total filings is carried out by aggregating forecasts that are made by the joint cluster planners.

Simple trend analyses can be applied separately to each of the 14 data series, in order to produce 14 forecasts. But once these are added up, the resulting totals do not normally agree with the same trend analysis carried out once on the total number of applications. The difference can be proportionally re-distributed to each group, to obtain a corrected series of forecasts. Going in the other direction, semi-automatic procedures can be used to separate forecasts of total filings into additive components that correspond to each of the 14 joint cluster totals. One method consists of trend analysis of the proportions of filings that take place in each joint cluster. This takes account of the interdependencies of the series and also presents an advantage in that the extrapolations comply with the total when they are summed up.

To see this, let

m = the number of joint clusters, $j = 1, ..., m$,
n = number of time points for the model, $i = 1, ..., n$,
y_{ij} = number of filings in joint cluster j at time t_i,
w_{ij} = proportion of filings in joint cluster j at time t_i,
w_{jav} = average value of w_{ij} over the n time points, and
t_{av} = the average value of time t_i over the data set for the model.

The following argument shows that the sum of the predicted values for w_{ij} over joint clusters is always 1.

Separate straight line regression relationships are fitted to the observed joint cluster proportions w_{ij} with respect to time for each joint cluster j. Elementary statistical theory (e.g. Draper and Smith 1981, p. 17) shows that the predicted values, w_{ijpred}, are given from the model as

$$w_{ijpred} = intercept_j + slope_j \times t_i$$

$$= [w_{jav} - (b_{jpred} \times t_{av})] + (b_{jpred} \times t_i) ,$$

---

[1] See Chap. 7, Sect. 4.2 for the names of the EPO joint clusters.

where

$$b_{jpred} = \frac{\sum_{i=1}^{n} (t_i - t_{av})(y_{ij} - y_{jav})}{\sum_{i=1}^{n} (t_i - t_{av})^2}$$

The key observation is that the sum of the predicted slopes across joint clusters is always equal to 0. Expansion of the terms in the top line of the second equality below shows that this is so.

$$\sum_{j=1}^{m} b_{jpred} = \frac{\sum_{j=1}^{m} \sum_{i=1}^{n} (t_i - t_{av})(y_{ij} - y_{jav})}{\sum_{i=1}^{n} (t_i - t_{av})^2}$$

$$= \frac{\sum_{i=1}^{n} \sum_{j=1}^{m} (t_i - t_{av})(y_{ij} - y_{jav})}{\sum_{i=1}^{n} (t_i - t_{av})^2} = 0$$

Therefore, the sum of predicted proportions can be obtained for any time point t_k, where t_k does not have to be equivalent to one of the input t_i values, as

$$\sum_{j=1}^{m} w_{ikpred} = \sum_{j=1}^{m} [w_{jav} - (0 \times t_{av})] + (0 \times t_k) = 1$$

This demonstrates the result.

Forecasts for joint cluster totals can also be made from the applicant panel survey. But, since the joint clusters are defined at EPO only, it is difficult to apply the transfer approach to joint clusters in the absence of information on first filings in other patent offices by corresponding definitions. Another problem is that many larger applicants make filings under more than one joint cluster. A way forward on this could be to form a concordance mapping of IPC codes to joint clusters, rather like the concordance between IPC codes and NACE/ISIC codes that was discussed in Chap. 5.

Contrary to most of the other data sets that are available on flows of subsequent filings between offices and systems, patent family data allow for counting of the numbers of priority forming references to patents originating at each office according to IPC codes. Therefore, it is theoretically possible to break down patent family counts to joint cluster equivalents using such a concordance table. However, the available patent family databases depend on patent publications and are incomplete for the most recent years.

It should also be borne in mind that the joint cluster structure at EPO was created for administrative as well as technical reasons. The groupings were chosen in order to make 14 sets of examiners of approximately equal size. IPC classes corresponding to joint clusters and the corresponding sets of examiners are moved between clusters from time to time to achieve balance. Taking this into account, it could be that an effective practical forecast for future filings per joint cluster may be obtained by first making a forecast for the number of total filings and then just dividing by 14.

2.5 Annual forecasting exercise

The forecasting operation is done as the first stage of the preparation cycle for the annual EPO Budget. Fig. 9.1 shows the various stages in the cycle. Five interrelated zones are indicated in this diagram.

1. Forecasts by various methods are prepared for a round table meeting, that takes place each year in January between the forecasting team and planning officers from various EPO departments, including representatives from all the joint clusters. The outcome of this meeting is a set of consensus forecasts for the current year and the following five years. The forecasts are the main part of a basic assumptions (BA) document for planning and consideration by the EPO's internal management committee.
2. After approval or amendment, the revised document is discussed in May by the budget and finance committee (BFC) of the administrative council (AC), and in June by the AC itself.
3. The EPO needs to forecast workload requirements for many downstream operations that take place during the granting procedure after patent applications have been initially filed. An example of one of these

downstream quantities is the number of euro-PCT regional phase filings (euro-PCT-RP). Other quantities include: workload requirements for search and substantive examination, numbers of final actions of various types, grant publications, special searches, oppositions, appeals, etc. After the approval of the BA document by the management committee, the units responsible for the planning function cooperate to construct a detailed medium term business plan (MTBP). This includes the major downstream quantities, with a refined evaluation of the numbers of incoming and outgoing applications at each stage of the granting procedure and an evaluation of the manpower capacity required to reach the objectives set out by the EPO management and the AC. The MTBP is discussed further with the management committee and is presented to autumn meetings of the BFC and the AC.
4. In parallel, the finance department constructs a budget document that is consistent with the draft medium term business plan. The AC meeting in December formally decides whether to imbed the forecasts that were originally made in January into the EPO budget for the following year. The lag between forecasting and budget approval can cause problems if the perception of the patent application market changes in the meantime. However, the procedure is consistent with the international treaty requirements upon which the EPO is based. The forecasted scenarios can be fine-tuned according to circumstances, although there is a strong presumption to maintain the original headline forecasts unless clear discrepancies with the actual out-turn have become apparent.
5. Data collection, generation of alternative sets of forecasts, methodological research and discussions with counterparts in other patent offices take place throughout the year with less time constraints. The applicant panel survey is organised, executed and analysed during the course of the year.

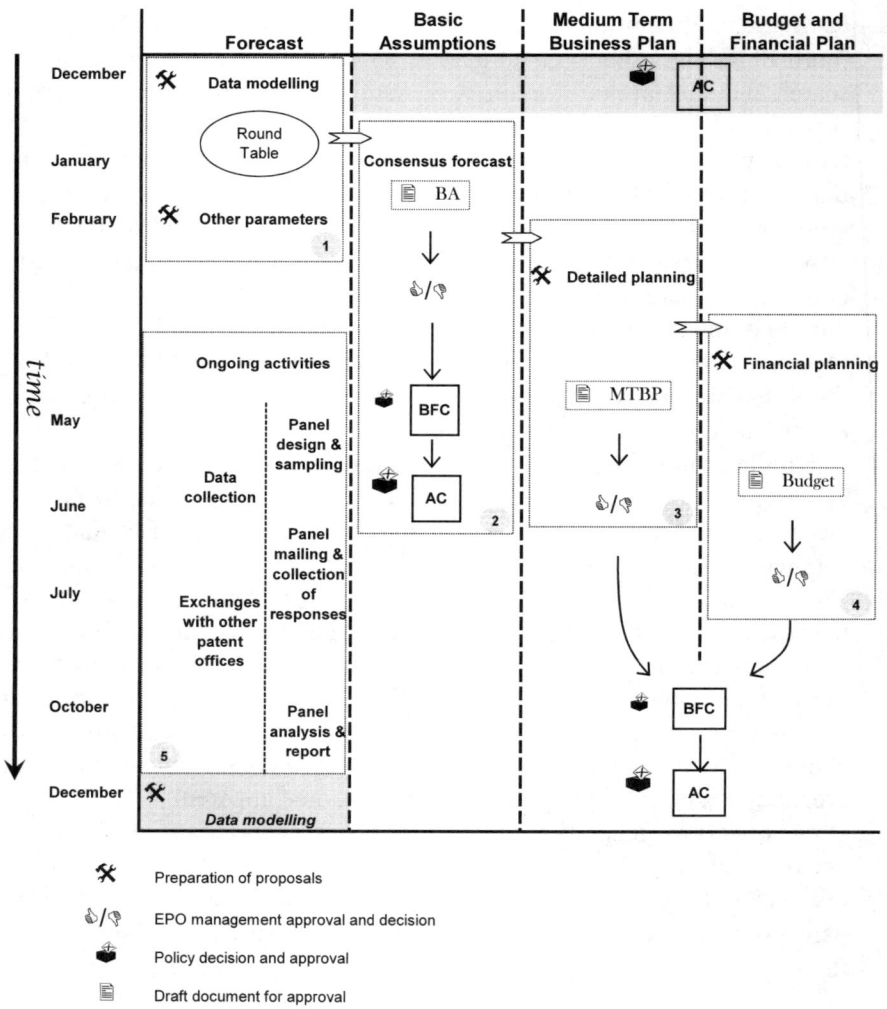

Fig. 9.1. EPO planning and budget cycle

2.6 Discussion

The existing approaches are applied over a variety of combinations of breakdowns of the total filings and time periods. This provides a bundle of forecast trends that indicate a general direction for future filings.

Agreement could be sought between the methods by calculating confidence limits on each projected scenario. There is certainly no need to work with 95% forecast intervals, because these are often too wide to be useful to decision makers, who may be happier to work with a lower probability coverage (Armstrong 2001). But, in practice, confidence intervals are not usually used because the straight line methods fail to fulfil basic modelling assumptions with regard to the underlying function that describes the trend and error structure. Also, it is not particularly useful to present confidence limits for forecast errors when the scenario is presented to the budgeteers, since it is difficult to disperse the error estimates to all the budgeted subitems in a meaningful way. It is rather more useful for the management committee to identify the direction of error that entails greater business risk, and to build in a planning margin to take account of this.

In order to understand the business planning function, it is important to realise that that the budget forming process requires input from many parts of the organisation about downstream items after the initial numbers of patent filings have been forecasted. A single scenario for the numbers of future filings should be chosen. It is inevitable that the out-turn will differ to some extent from the single scenario and business planning should take account of the relative risks of overshooting or undershooting the scenario. A limit of acceptable percentage forecast error is tacitly agreed between analysts and the budgeteers. Due to the heterogeneous nature of the European Patent system, this limit is a little higher for EPO than what might apply in the more monolithic systems at JPO and USPTO. By experience, an absolute forecast error of 6% for the following year is a reasonable expectation for EPO, although it should be accepted that this is an average level that may be overshot in exceptional circumstances.

The main aim of the forecasting exercise is to predict the total incoming filings workload in the following six years. Subsidiary aims are to produce forecasts for breakdowns, such as euro-direct vs. euro-PCT-IP filings; first filings vs. subsequent filings; and joint cluster breakdowns. Unfortunately these aims are not necessarily consistent. For the remainder of this chapter, we will concentrate on total filings and consider methods based on breakdowns mainly in order to comment on how these can improve the forecasting accuracy for total filings.

Since the scenario decision that underlies the budget is made some 11 or 12 months ahead of the start of the calendar year in which the budget begins to operate, the accuracy of the forecasts for filings in the current year and the immediately following year are crucial. Further out, accuracy is still important but the possibility remains to revise the scenario in the following annual exercise. For the latter years of the scenario, disjointed jumps in forecasts should be avoided because they cause difficulties in the long term planning of resource allocations.

For the derivation of forecasts for other workload items further down the chain, currently these are derived by the application of established ratios to the initial forecasts. The historical series of ratios indicates what proportion of the headline filings subsequently enter each downstream phase and with what delay. In practice, the ratios usually show slow trends that can be predicted without any undue difficulty. In some cases, a decision to alter the procedure in some way will affect these forecasts. As decisions of this nature are usually known prior to implementation, assumptions can be incorporated in the process. The ratios are updated at least annually, although there can be a delay in data availability which means that ratios for the previous calendar year are not available early enough to be used for the forecasting exercise. This approach has the advantage that the construction of projections necessarily remains consistent throughout the plan.

Patenting is a global phenomenon and the analysis of patent filing trends at other offices is relevant for projecting trends at the EPO. To this end, exchanges in forecasting methodologies and models have become more frequent between the main trilateral patent offices and WIPO. A trilateral statistical working group has been set up to facilitate these interchanges.

Our understanding of forecasting practice at the USPTO is that a forecast performance rating procedure is in use to select the model that performs best over recent history. This often excludes models with explanatory variables, in favour of the self determining approaches. The concomitant variables should however contain information that can enhance the forecasts under a good model. The degree of enhancement is however often only minor, as is shown in Chaps. 5, 6 and 7. Nevertheless, we believe that the successful incorporation of at least one additional explanatory variable is also helpful to improve the validity of the model specification, a factor that can differ from the degree of improvement of forecast performance.

We believe that at JPO a forecasting system is used that is broadly based on projecting linear trends of important components, such as types of filings and numbers of filings from different blocs. Downstream items from filings are also directly forecasted. Special considerations are then given to one off events, such as the reduction of a grace period between filing and request for examination from seven years to four years that occurred in 1999. At WIPO, a country wise modelling approach is in use, with the selection of appropriate models for the projection of PCT filings from each country according to statistical criteria.

A multi-organisation patent statistics task force has been set up, under the auspices of the OECD, in which forecasting and other matters are discussed within a wider group, including the trilateral patent offices, WIPO,

National Science Foundation of US and the European Commission (OECD 2005). This task force also considers the policy implications of patent statistics. The inherent interests of the patent offices in forecasting their workload complements the wider policy implementation of forecasting technology trends in society. Academic economists become involved as consultants to the task force members, or in other fora such as in the interdisciplinary association on European Policy for Intellectual Property (EPIP 2005).

3 Research programme recommendations

The research concentrated mainly on regression analyses of the time series structure of the macroeconomic patent filings data. Chapter 8 was slightly different in that it analysed the patenting behaviour of individual applicants.

The time series approaches consider not only the self determining properties of the patent series (Meade, Dikta), but also the prognostic ability of associated economic variables such as R&D and/or GDP (Blind, Park, Dikta) and other factors (Blind). These relationships are not always clear cut and reproducible between countries at the macroeconomic level. A goal of the research programme was to establish whether forecasts can be improved by making use of the concomitant information and, if so, how is it possible to use such information in an optimal way.

3.1 A research programme for improving forecasts of patent filings (Chap. 2 by Dietmar Harhoff)

An earlier version of this review of the aims of the programme was given as information for potential participants in 2000. The breakdown of the structure into the five modules A–E (Fig. 2.2) represents a useful typology of the spectrum of possible approaches in this area. In fact, most of Chaps. 4 to 8 cover more than one module, which reflects the breadth of the research projects that were carried out.

In Sect. 5, of Chap. 2, Harhoff recommends that most data should be provided to the researchers by the EPO, and we have followed this suggestion. There was a problem of timeliness, since the programme ran for approximately three years and data were supplied at different moments. Also some items in the databases are incomplete for a considerable time after the fact. In practice, real forecasts must also be made with the latest available information and updated as often as possible.

In Sect. 4.2, Harhoff discusses ways that the annual EPO applicant panel survey can be improved. Some of the suggested improvements have already been made. On the issue of questionnaire complexity, every effort is made to restrict the questionnaire to no more than four pages and telephone support is given to the respondents to help them to fill out the answers. They are encouraged to reply by internet, fax, letter or telephone, though not many have taken up the internet option. Regarding incentives to respond, we are grateful that the respondents feel that it is in the interest of the community at large that EPO receives the information that it needs for resource planning without reward. Regarding bounded influence estimators, the use of these is under consideration and Winsorised estimation of the year to year growth indices for filings has been done for the first time in the applicant panel survey of 2005 (EPO 2006). On the possibility of joint survey work with JPO and USPTO, this issue has been addressed but is difficult to achieve. The necessity to respect the confidentiality of clients' responses to the surveys at each office creates constraints that prevent all parties from appointing a single global consultant. Instead, the overall survey results and interpretations are shared between offices as far as possible. We also seek to include a complementary minimal subset of questions on each survey, in order to improve the comparability of the results.

In Sect. 6, the suggestion for a "researcher in residence" has been taken up by the EPO with the appointment of a chief economist on a rotating basis. A research competition has not yet been instituted, but the recommendation in Sect. 6.2 for sponsored conferences has been fulfilled to some extent with regard to patent statistics and economics in general. For example, EPO hosted the first EPIP Conference in Munich (EPIP 2003) and cosponsored the 2nd WIPO/OECD workshop on patent statistics (WIPO 2004). At both of these events, papers on forecasting patents were presented. EPO certainly wishes to encourage further work in these areas.

3.2 From theory to time series (Chap. 3 by Peter Hingley and Walter Park).

Aspects of this chapter are discussed in Sect. 3.6 below, in conjunction with a review of Chap. 7.

3.3 An assessment of the comparative accuracy of time series forecasts of patent filings: the benefits of disaggregation in space or time (Chap. 4 by Nigel Meade)

This chapter uses self-determining methods of time series analysis on the patent series, exploring in particular the ARIMA and VAR (dynamic linear model) approaches. It concentrates on a training set (filings up to 1994, 1995, 1996, 1997, 1998, 1999), followed by data horizons of up to four years on which to assess the methods, with formal testing carried out for the one year ahead forecasts. The modelling methodologies are explained and then techniques for statistical assessment of forecast performance are introduced. The literature review in this chapter includes references to applications of forecasting techniques in other fields that have experienced the benefits of globalisation, particularly the tourism industry.

The question is explored of what can be gained in terms of forecasting accuracy by disaggregating the patent series according to blocs of geographical origin or by the technical fields of the applications. A subsidiary question is whether monthly forecasting is a more accurate way to assess future filings than annualised forecasting.

Self-determining approaches have limitations in that causative factors in the development process are ignored in favour of a consideration of only the inherent trend of the target data set itself. The self determining system could perhaps be considered as a kind of modelling null hypothesis, against which models that contain prognostic variables can be tested to see whether they give a significant improvement in forecasting accuracy.

However, there are some arguments that tend to favour a self determining time series approach:

- The patent series may consist of underlying trends which form an equilibrium that can be re-established after periods of exuberance or depression, themselves caused by world wide economic trends and to levels of patent office fees and other costs. Unlike the case of many economic variables, for EPO filings the trend has been upwards with only localised perturbations. There was a slight downturn in 1991 when there was a nearly simultaneous fee increase and an economic shock due to the effects surrounding the first Gulf War. In 1997, there was a short term surge in applications after the time for payment of country designation fees was put back from the moment of application to the time of request for substantive examination. There was another small downturn in 2001 which seems to have been due to the piercing of the stock market bubble for many technology based firms in 2000. These kinds of events can not be easily predicted.

- The patent series are affected by other measurable macroeconomic series, but the nature of the dependence is not clear and is also heterogeneous over time. It can be argued that, if the relationship was sufficiently clear to have been of critical importance to modelling, it would have been described and routinely applied long before now.

The techniques that Meade describes are relevant and worthy of consideration in developing further modelling approaches. Exponential smoothing is a technique that can be valuable for short term modelling, and which is easily used for monthly data for particular technical areas, joint cluster groupings or more specialised groupings of examiners (*pensa*) within the joint clusters. The state space approach has been identified as a good way to generalise exponential smoothing methods (Hyndman et al. 2002). Dynamic linear modelling is a variety of the state space approach, which is itself a flexible framework for selecting and comparing models. State space models can also be used with prognostic variables, and this technique can avoid some of the pitfalls of ARIMA based transfer function modelling (Durbin and Koopman 2001).

Meade suggests that there are at most only limited advantages to be gained from the disaggregation of series by technical areas or blocs of origin. This finding should be treated with caution. No particular benefit was identified for breakdowns into 9 technical areas that were indicated by the first digit of the IPC code (including a residual category). It could be that this level of breakdown is not detailed enough. Inspection of patenting trends since 1990 suggests that a more meaningful breakdown could be made between areas of rapid development (e.g. pharmaceuticals, computers) and those with a more stable or stagnant trend (e.g. inorganic chemistry, construction). A purist might argue that this puts the cart before the horse – a split into high and low growth groups presupposes the result of the forecasts before making them. But it can make sense to do this as long as future developments can be expected to mimic the past. After the stock market experiences of the late 1990s, it might be better to rename these groupings *high risk* and *low risk*, rather than *high growth* and *stable growth*, and so define them in terms of differences in the volatility in year to year growth rates. It is a testament to the steady success of the patenting system in recent history that *high growth* seems to have been synonymous with *high risk* over most of the period in question.

The experience at the EPO of forecasting by disaggregation is that it is certainly important to model euro-direct and euro-PCT-IP filings separately. The numbers of euro-PCT-IP filings amplified greatly after they became popular amongst US based applicants in the late 1980s and now constitute up to about 70% of overall filings. This itself creates a modelling problem, since part of the process has been a steady substitution of euro-PCT-IP for euro-direct. We assume that there is probably an absolute upper threshold around 80% of PCT usage for EPO filings from most countries of origin, though achievement of this proportion occurs at different times for each country. It may be appropriate to use saturating or sigmoidal models for the development of the euro-PCT-IP proportion of overall filings over time.

It is certainly important to take the countries of origin of the filings into account, since economic and legal circumstances and rates of development vary considerably between them. At EPO, we regularly fit separate models to the main blocs of origin (EPC contracting states, Japan, US and Other countries). It has been argued that this sometimes gives a demonstrable advantage in terms of forecasting accuracy and sometimes not (e.g. applicant panel report 2003 vs applicant panel report 2006, EPO 2006). A breakdown by blocs is also effective in the transfer modelling approach. However it is also possible that we have not yet gone far enough in the consideration of geographical breakdowns – the models developed by Park in Chap. 7 are based on about 30 individual countries and have been quite successful.

With regard to industrial areas, at the EPO there is a particular need to model filings at the level of the 14 joint cluster groupings. Here we are not giving consideration to the methods most appropriate for forecasting such breakdowns *per se*, so it can only be said that no particular advantage in forecasting the total has yet become clear from the disaggregated approach, and in this matter we concur with Meade.

Regarding analysis of monthly data for forecasting, Meade favours this under some circumstances. But practical considerations of data availability have to be taken into account. In the live forecasting setting, the filing data for recent months may be particularly tentative and, although this also applies to the most recent running annual total up to the previous month, or to the annual filing total soon after the end of a calendar year, the coefficient of variation due to uncertainty is less for recent annual counts than for recent monthly counts.

A criticism of the approach taken in Chap. 4, and in some of the other chapters as well, is that it is not necessarily appropriate to draw permanent conclusions from a study based on a restricted set of training and test years. Also it is debatable whether it is good enough to assess relative forecasting accuracy based only on one year ahead forecasts.

Meade did at least try out a small selection of alternative training sets. He chose a prediction zone of one year ahead forecasts for the years 1995 to 2000, that is based on actual filings data sets starting in 1979 and ending one year before the year to be forecasted. The years 1995 to 2000 marked a period of steady high growth in European patent filings. It is difficult to make firm conclusions on the benefits of disaggregation without further investigation of potentially less benevolent prediction regions. The models developed in Chaps. 5 and 6 have been developed on a one off basis using training data up to 2000 (Dikta) or 2001 (Blind), and tested on the years thereafter. The more recent years of 2002 and beyond provide a more difficult prediction zone than 1995 to 2001, because the filings development was flat in 2002 and 2003, followed by a recovery thereafter (see Chap. 1, Fig. 1.1). Meade does in fact forecast out as far as 2003, on a four year ahead basis, with training data up to 1999. The forecast for 2003 in Fig. 4.5 is almost 180 000 filings, which is rather higher than the observed out-turn (167 000). So these models have not handled the period since 2000 very well, although no other self determining model could have been expected to do much better.

3.4 Driving forces of patent applications at the European Patent Office: a sectoral approach (Chap. 5 by Knut Blind)

In this contribution, patent filings data series were created for industry/country combinations using a concordance structure that was developed between patent classifications (IPC) and industrial area codings (ISIC and NACE). Time series regression analyses were then carried out for each series against a number of prognostic variables.

This study goes beyond country analyses, in order to consider sectors and combinations of countries and sectors. For this purpose, it is necessary to allocate patent filings to industrial sectors within countries. A limitation is that data are distributed according to patent classification systems, which are defined in terms of scientific and technological classifications, like the IPC, while economic performance indicators are based on rather different economic classification systems, which are based on areas of industrial activity, such as ISIC and NACE.

In Sect. 2 of Chap. 5, the construction of a concordance matrix between IPC and ISIC is explained. This aspect of the work is not of primary interest for the purpose of improving forecasts and highlights some inherent difficulties with the methodology, although it is of potential use in other domains of patent data analysis. Firstly, the connection between IPC and ISIC relies only on work by experts to allocate each of the 625 IPC subclasses to one of the 44 industrial categories. This is based on a sample of only 3 000 enterprises – an impressive enough number in itself but nowhere near the whole patent universe. There are therefore some deficiencies in the resulting concordance matrix (only one sector for one company, only 12 countries under consideration). This was in fact just the most recent in a series of attempts to link these two types of classifications. There is not yet a widely recognised linkage between IPC and ISIC (or NACE), which could serve all econometric analyses of patent filings. See WIPO (2003) for further discussions on concordances.

Some aspects are promising, since the concordance matrix can be assumed to remain stable for several years. However, the exercise assumes that patent activity results mainly from the manufacturing areas and leaves aside the non-manufacturing sectors by presupposing that they have much less propensity to patent. We doubt this assertion and suggest that it requires further investigation. Also, patenting by individuals is ignored. In Chap. 8, Frietsch estimates that 28% of applicants are individual free inventors (discussed in Sect. 3.7 below). This indicates that the contribution of individual inventors to patent filings is not negligible.

The results are highly country and sector dependent, with implications for the stability of the concordance matrix. Time lags and signs of effects also vary between countries and sectors. If the EPO were to implement this particular forecasting methodology, it would be necessary to expand the model to more than 12 countries. Few of the advantages of using a simplified model will apply here.

Blind studied the dependency of patent filings on more than just GDP and national R&D expenditures. The need to look for additional economic factors is justified by the results of an earlier study that demonstrated that the relationships between R&D expenditures and patent filings have become looser in recent years, particularly in Germany. The growing tendency towards globalisation led Blind to look for networking and international market structure indicators. In particular, four exogenous indicators are used as explanatory variables to improve the quality of forecasts of future patent applications at the EPO. These are: R&D, industrial standards, value added and exports.

The form of the R&D expenditure data that was used is ANBERD in PPP US$ (OECD MSTI).

Section 3 gives the results of the empirical study that was conducted on the selected data. The analysis is conducted in three stages. Firstly, the total model is tested to evaluate the influence of each of the four explanatory variables. Then a similar model is applied to each country. Finally, the models are applied to each sector. In each case, the time lags of the four variables are analysed.

To prepare the time series for analysis, inclusion of a trend variable has the effect of removing the directionality of the data. This is an alternative to the method of differencing that was used by Meade. However, in the light of the consideration of Box-Jenkins style time series effects by Dikta

(Chap. 6), it seems that the use of such a trend variable may not be enough to remove all the coincident causality effects between input and output variables.

On studying the time lag between R&D expenditures and patenting, Blind reports lags of one to two years for cases where R&D has a positive effect. Time lags are discussed further in Sect. 3.5 below.

In general, when relationships between patent series and explanatory variables could be identified, they were in the expected direction. However, a great deal of heterogeneity was discovered between countries and industries, and this causes difficulties of implementation for forecasting purposes. The explanation that is suggested for this is that the nature of the relationship between R&D expenditures and patenting has changed since the early 1990s, with a higher propensity to patent per unit of R&D, due to enthusiasm for the patent system and the attractiveness of high-patenting firms to investors. Some support for this argument appears in Chap. 6, Fig. 6.3.

In most of the combinations considered, the measured effects are significant and lags can be identified. Together with the dependencies of the effects on the individual combinations of countries and sectors, this suggests that within countries there exist dominant sector(s) having a strong influence of the results. A second consequence is that the effect of one explanatory variable may be significant but be in a different direction between countries or between sectors. The analysis should be deepened to consider more carefully the interaction effects of the four explanatory variables. Blind recommends the use of weighted least square methodologies and multivariate approaches to better account for cross-equation heteroscedasticity.

Industrial standards emerge as a possible leading indicator of patenting that should be studied further. It is a logical finding that these act as good predictors only for the more high technology sectors. But it seems rather strange that a positive relationship between R&D and patenting only appears for low-technology sectors.

The divergent results that have been found from so many time series runs are not especially useful for our main aim to forecast total EPO filings on a six year time frame. We can accept that the underlying economic drivers are heterogeneous. But it might be more useful to have an abstraction of the economic system that is consistent with it, yet capable of producing the required forecasts in a quasi-systematic way. As Harhoff mentions in Sect. 4.4 of Chap. 2, a better alternative is to aggregate industries into about five or six major groups for forecasting purposes. Blind's method is a glorification of the bottom-up approach – fascinating to look at the minutiae and useful to identify developments in particular fields of interest, but unlikely to provide a comprehensive model for the whole system without specifically accounting for large numbers of special case scenarios. We are unlikely to want to pursue this approach further for central EPO planning, but do recommend it as a possibility to joint cluster and *pensum* level planners.

The study throws some light on the difficulties of merging patent data with economic performance indicators. Furthermore it gives an alternative to more elaborate models, by offering simple regression functions to be replicated over combinations of countries and sectors.

Regarding other alternative explanatory variables to R&D expenditures, one possibility that Blind did not consider is the numbers of researchers in a country. This is something for further investigation, particularly considering the fact that there are too many ways to measure R&D expenditures and discount them over time before fitting regression models.

The time series aspects tackled in Chap. 5 could be extended to the micro level, to see whether or not similarly formulated models can be fitted successfully to individual companies. Can a scale-up be made from the individual company behaviours to whole sectors? It certainly seems that the sector based approach to macro forecasting is problematic, and an aggregated micro approach might therefore be the key to understanding sectoral behaviour. This might be done best by examining the growth rates of patents and predictors rather than absolute levels. Sources of variability and heterogeneity between companies will need to be carefully described. (See Sect. 3.7 below)

3.5 Time series methods to forecast patent filings (Chap. 6 by Gerhard Dikta)

The starting point for this approach to the analysis of patent data is the well established time series theory for business forecasting that was promoted by Box and Jenkins (1976). The methods involve autoregressive moving average models (ARMA) on differenced data, auto distributed lag models (ADL) and vector autoregressive models (VAR). ARMA on differenced data is formally equivalent to ARIMA analysis as introduced by Meade in Chap. 4. AR(I)MA is used by Meade to describe the patent series in a self determining manner, but, like Blind in Chap. 5, Dikta goes further by investigating how information from concomitant economic series can be used to improve the forecasts. This involves a careful analysis of the time series modelling properties of a system including one or two candidate predictive series. Blind was more experimental, by including several predictor variables in the system and worrying less about the structure of the data.

Dikta outlines a series of fundamental steps that are required to prepare the time series via transformations before applying the models. Firstly, an appropriate Box-Cox transformation is applied where necessary, such as taking logarithms to achieve homoscedasticity. The patent series data are then differenced (usually twice) to ensure stationarity. The complexity of the tests and transformations is such that only fairly straightforward linear regression models remain easily available for the data fitting phase. The methods are first applied to the patent series themselves in order to give a self-determining model that is comparable to the approach of Chap. 4. Then similar methods are used to prepare the series of predictors (R&D, GDP) as input series. These series are prewhitened in order to ensure that all accidental covariation is weeded out, and then a cross correlation analysis is carried out to find the appropriate lag between predictor and patents. After this, a linear model is fitted between the transformed data sets. Finally, the results in terms of forecast estimates are transformed back to the original metric of the data.

This approach is fairly rigorous and reduces the danger that apparent causative effects of the predictors are caused by spurious correlations, as may happen if both predictor and patent series increase over time due to external effects without a common cause. On the other hand, estimation errors might be increased after applying the transformations and can lead to wide confidence limits for the forecasts. There seems to be a balance to be achieved between ensuring that the data behave properly before analysis and using a relevant economic model.

In addition to looking at the total filings series, Dikta has taken a blocwise approach to account for regional variations in patenting behaviour towards the EPO. Residents of EPC contracting states, Japan and US are considered separately. It would have been difficult to apply the rigorous data checking methods with a more detailed breakdown by countries of the applicants.

For this research project, the form of the R&D data that is selected is BERD at US$ PPP (OECD MSTI, with currency conversions from national currencies made by the author).

An ARMA based transfer function model approach is used to prewhiten the data and to identify the most likely lag between R&D expenditures and patenting. The evidence suggests a lag of five years between R&D and European patent filings. This lag seems rather long – in particular there is an apparent conflict with Blind's results in Chap. 5 (approximately one year) and with survey data collected by the EPO, in which respondents estimated an average (mean) lag between R&D expenditure and first patent filings of only 14 months (EPO 2006). On the other hand, Dikta's result is fairly consistent with an average lag of 3.7 years from R&D expenditure to first filings in EPC contracting states that was reported by Hingley (1997).

How can the discrepancy in lags of almost four years between microeconomic and macroeconomic approaches be explored? Firstly, it should be noted that most EPO filings are subsequent filings that take place after the first patent filing for an invention – this accounts for one year of the difference with the survey data. Secondly, it seems plausible to imagine that the research development time for a particular patentable invention may be short, yet the building of whole research departments takes up most R&D funds and involves a longer investment time that results in bundles of patent applications. The matter is unclear and worthy of further investigation.

Given the rigour applied to the analytic assumptions underlying the cross-correlation model used to establish the five year lag, this result should be useful outside the confines of this particular project. It is also fortuitous that a long lag can justify the construction of projections several years forward, using the R&D information that is currently available.

In the second part of Chap. 5, Dikta uses an ADL approach that considers developments in the various blocs and correlations with the prognostic variables R&D and GDP. Two models are suggested. In the first model there is a one period autoregressive term, a four period lagged R&D component and a simultaneous GDP component. Another model retains the one period autoregressive term, but replaces the lagged R&D term with GDP at various lags as a proxy variable. This has the advantage that there are several readily available forecasts for GDP that can be used in the patent forecasting operation. It removes the need to work with the more messy R&D data and allows patent forecasts to be made further into the future without first having to make R&D forecasts. This is an important practical consideration.

Difficulties with Dikta's approach include the requirements for specialised software and trained analysts, as well as the fact that the method is rather formalistic. As we remarked above in Sect. 3.3, the proponents of state space models suggest that their method avoids some of the prescriptive difficulties of the Box-Jenkins approach, so a state space method involving predictive variables is something else that could be tried. The theory that underlies Dikta's technique is appropriate for studying physical processes and may be rather too sophisticated for economic systems that depend on various changing underlying factors. For forecasting purposes, we feel that lagged associations of prognostic variables with patents are enough. If we can find that lagged R&D or GDP generally moves in step with patent filings, this is sufficient for building a forecasting model. It may not be necessary to remove all the trend effects systematically by multiple differencing before performing an analysis based on the assumption of a relationship at a certain lag distance between the series. Knowledge of causation is better for predicting trend breaks in a complete model of the process, but, if the variables associate through history, we can believe that they will associate in the future and can use them for planning purposes.

In Dikta's third approach, a multivariate VAR model system is developed that involves R&D and GDP as well as total filings. Both unconditional and conditional variants are discussed. This is an extension to the use of VAR that was discussed in Chap. 4 for describing the self-determining process. The unconditional approach also has the interesting property that it provides forecasts for GDP and R&D as a side-product. However, the shortness of the available series restricts attention to lags no greater than two years. Therefore, Dikta has pragmatically imposed a 5 year lag on the R&D series before fitting the conditional VAR models, making use of the finding about the lag that was obtained using the ARMA based transfer function model approach. At this point it was clearly advisable to be pragmatic and to slightly relax the degree of theoretical rigour in order to obtain good results.

The VAR approach has potential for further development. For example, it could be tried out for modelling euro-direct fillings and euro-PCT-IP together, or for global models for filings from several patent offices together.

Coming back to the issue of whether it is better or not to aggregate forecasts that are broken down by blocs of origin, we note that Dikta's conclusions (in Table 6.15) indicate that it is perfectly feasible to apply his approaches on the total filings series without breaking down by blocs. Consideration of blocs, countries or technical areas may be safer, at least to make comparisons, but the aggregation procedure brings in questions of correlations between series that need to be addressed even when the disaggregated forecasts are better than those for the total. This problem is reduced to some extent in the VAR approach, where correlations between disaggregated series can be automatically handled. We look forward to making further experiments with VAR.

We appreciate the practical insights that this author has brought to the forecasting problem, and that the study is highly consistent with the stated aims of the research programme. However, Dikta's approach gives wide confidence limits on forecasts. Although this is probably realistic – since the future is indeed only partially knowable – rigorous simulations would be advisable to check for validity under a variety of data generating mechanisms. In comparison to Dikta's approach, Park's method in Chap. 7 suggests that a simpler econometric method using multiple applicant countries without differencing also works well. It should also be noted that the inclusion of the concomitant variables (GDP and R&D) leads to only a small advantage over the self-determining system. This is reasonable if return-to-equilibrium processes take place within the patent series and if other unmeasured variables are influencing the development, in addition to the pair of prospective variables that have been selected. It would be interesting to include elements of error correction models in this approach, along the lines that have been suggested in Chap. 3.

3.6 International patenting at the European Patent Office: aggregate, sectoral and family filings (Chap. 7 by Walter Park)

This chapter describes another regression modelling approach that includes prognostic variables for forecasting EPO filings. In Chap. 6, Dikta concentrated on the time series structure of such data series. By comparison, Park adopts a more explicitly econometric approach that is broadly connected to the theoretical model of the patenting process in Sect. 2 of Chap. 3. This theoretical model could be of some general interest for deriving forecasting models that take account of drivers for patenting, including the characteristics of patents as options on the exploitation of technological innovations. This borders on methods that can also be used for assessing the values of patents.

Because the modelling methodology in Chap. 7 is fairly simple, it is practical to apply it over a larger number of individual applicant countries, rather than just the four main blocs of origin. Data are transformed by dividing variables by the workforce size per country and then taking logarithms. Park then uses linear regression models that incorporate effects of concomitant variables (R&D etc.) and lagged values of the patent series themselves. The models are fitted on the available historical data with common parameters between countries to generate predictions for future years. Finally the predictions are back-transformed and summed over the countries in order to form the filings forecasts.

For this research project, the form of the R&D data that is selected is GERD for most analyses and BERD for sectoral analyses. The data were available in real PPP dollars from the OECD and from UNESCO's Statistical Yearbook.

This method allows for easy experimentation by trial and error to look for the best predictors. Experimentation is also possible on different types of patent data. Park uses a hierarchical approach, beginning with an analysis of aggregated data that are split only by combinations of filing types (first filings, subsequent filings) and filing products (euro-direct, euro-PCT-IP). Then, he considers sectoral filings in the form of breakdowns into five manageable technology groups that are made from accumulations of the 14 EPO joint clusters. This conforms to one of Harhoff's suggestions (Chap. 2, Sect. 4.4). Finally, Park considers making forecasts for patent families that lead to filings at the EPO, thus allowing for the mapping of economic effects of internationalisation from country of origin to other countries.

At all stages of the research, assessments of forecasting error have been made by comparing the predicted values with the actual out-turns. Park uses root mean square proportion error as the statistic for this and we comment further on model comparisons in Sect. 5. However, the assessments have been done only on the basis of models that have been fitted on data up to 1996. A wider variety of training sets and test periods would be advisable.

There is a philosophical difference between this method and the more mechanistic approach of Dikta in Chap. 6. It would be beneficial to compare both techniques in terms of forecasting accuracy. Methods to calculate confidence limits on forecasts are required for both methods. At present, no formula is presented for the Park approach, though it might be expected that the confidence limits will be narrower than for the Dikta approach, which requires more rigorous checking and transformation of the data. This does not necessarily make the Dikta approach any worse. If Park's method seems better because it is easier to apply, an inflation of its confidence limits might be advisable to take account of the degree of pragmatism involved. However, forecasting accuracy is not properly measured only in terms of widths of calculated confidence limits on the forecasts. Rather it is the degree to which the forecasts correspond to the later reality that determines the usefulness of the method. Dikta's limits might turn out to be more accurate in terms of the historical performance of the forecasts against actual out-turns.

The hierarchical approach has perhaps not been carried out to the fullest extent. No special conclusions have yet been drawn as to the comparisons of the resulting forecasts to determine which level of disaggregation is most appropriate. Does the split by joint cluster or groups of clusters give better forecasts for the total than the overall forecast? Forecasting was also done via breakdowns done by product type (euro-direct, euro-PCT-IP; first filings, subsequent filings). Therefore product type forecasting by joint clusters and technology groups must be done before a proper answer can be given. This aspect was perhaps covered better in the more limited exercise of Chap. 4.

As far as patent families are concerned, we believe that these data hold the potential for a better structural model of the international patenting process. Unfortunately, the data are hampered by being restricted to published applications. They are not accurately counted for several years after the fact and this diminishes their usefulness for forecasting. We hope that these data can be improved in the future, either by including more up-to-date counts that are based on as-yet unpublished applications, or by establishing the relationship between families data and live filings counts data more perfectly, so that the missing families data for recent years can be now-casted from the filings data themselves (Hingley and Nicolas 1999, Dernis and Khan 2004). Good quality patent filing data may hold the key to the improvement of forecasting methods in a world of continuing internationalization.

Is it better to account for bloc/country of origin effects by fitting separate models to each bloc/country, or should a combined analysis always be carried out? The separate approach allows for different models to be fitted in each case to maximise goodness of fit, while a common approach allows for common parameter fits where no significant differences exist between blocs/countries. However, the inherent assumption of Park of common slopes and intercepts for each country may be too simplistic. The pace of patenting differs by countries. More particularly, the pace of international patenting at the EPO differs by country of origin, with newly developing countries increasing their year-to-year applications much more than the norm. Therefore, we are looking to examine the benefits of fitting separate slopes and intercepts to the country by country data, as well as evaluating the possible advantages of separately forecasting by product type or technical classification. Preliminary results of a country by country breakdown suggest that a distinction can possibly be made between two main groups of applicant countries. On the one hand, the larger and more established participant countries, such as Japan, Germany and US, have high intercepts and low but positive slopes to indicate continuing steady growth. On the other hand, smaller countries and players that are establishing themselves in the system tend to have lower intercepts and high positive slopes, which indicates the potential for rapid growth.

Regarding the applicants' choice between the euro-direct and euro-PCT-IP filing routes, Park separately estimated filings by these routes, but has not considered whether the total is better estimated by treating one route as a substitute for the other, with implications for the covariation between the corresponding time series. Some account should be taken of the sociological trend away from euro-direct towards usage of the PCT system. One way that this could be done is to model all subsequent EPO filings as the main variable, then within this allow slowly changing trends in proportions of euro-direct vs euro-PCT-IP. A robustified Parzen window type regression approach might be useful for this (Insightful 2001, p. 122). As a place of first filing, the EPO is in competition with EPC national offices, so national office priority filing data should be included. The process could also be informed by flow/transfer effects involving patent families modelling analysis.

Logic suggests that R&D investments have an effect on first filings, and then the transmission to subsequent EPO filings involves globalisation factors. Thus, a more detailed structural approach tends to downplay the relevance of a model for total filings based on direct effects of R&D investments. None of the researchers have delved too deeply into this issue. Park concludes that R&D best explains subsequent filings. EPO subsequent filings are the main component of interest for our forecasting operation, and the relative stable growth over time of these data may explain why R&D predicts these best

Another issue is the effect of patent fees on numbers of applications, and the extent to which a fee parameter can improve the goodness of fit of a forecasting model. Consideration of fees and associated costs of patenting can be approached within the framework of models such as those introduced in Chap. 3, Sect. 2. Park has reported elsewhere that the analysis of historical data suggests that fees are a significant predictor, but do not give additional significance after taking account of lagged R&D expenditures and other terms. If the purpose of an exercise is to simulate the effect of proposed fee changes on filings, then a model including a discounted fees term is appropriate. For general forecasting purposes, there seems to be no particular advantage to include such terms however. In our experience, sudden fee changes can lead to temporary displacements in the filings series, but these do not generally seem to have much effect on long term filings trends.

On the issue of how best to incorporate R&D filings series into the models, in unpublished work Park experiments with a model that involves time depreciated stocks of R&D rather than raw R&D flows, and concludes that no particular advantage is to be gained from using stocks rather than flows. JPO has also experimented with forecasting models that include stocks of R&D expenditures and stocks of patent applications.

Park's approach has potential as a routine forecasting tool for econometric modelling. However, as in Chap. 4, the models developed for total filings and for sectoral filings have been estimated on a data set that gives a relatively easy forecast zone when there was a continuous period of growth. In Sect. 4 below, we give an account of an attempt to fit a variant of Park's model to filings data up to 2003, where we find that the system's forecasting power remains acceptable, though not at all perfect.

Improving forecasting methods 219

3.7 Micro data for macro effects (Chap. 8 by Rainer Frietsch)

This is a pioneering effort to explore the sources of information that are available for forecasting at the microeconomic level on the activities of the companies that apply for European patents. Data sets are formed of economic variables per company, using web based information services and data on patent filings from EPO databases. There have previously been several studies that show that the relationships between firms are quite strong at the cross sectional level, explaining differences between number of patents applied for or granted to firms by differences in their R&D efforts. However, the relation appears much weaker, though statistically still significant, in the time-series dimension, when explaining changes in the number of patents for a firm by changes in its R&D efforts. Earlier reports suggest an elasticity of the number of patents to the accumulated stock of research capital of about 0.3 (Griliches 1990; Crepon and Duguet 1997; Crepon et al. 1998). Studies have also been performed which give insight into the use and function of patents on the firm (e.g. Bertin and Wyatt 1988; EPO 1994). But it seems to be quite difficult to bridge the micro- and macro- levels without making a leap of faith.

Firstly, Frietsch analyses a random sample of applicants in order to develop estimates of overall numbers of patent filings, and then descriptive statistics are presented for factors that correlate with patenting levels by companies. An additional logistic regression approach is taken to look at the question of whether applicants will file at all in future years. Then there is an analysis of the companies that spend the largest amounts of R&D in the world, including descriptive statistics and correlations of patenting with R&D expenditures. The question of continuity of applications was not considered for the largest cases, since these companies are unlikely to stop patenting entirely in the near term.

Within the random sample, Frietsch decided to concentrate on variables that could actually be extracted from the databases, particularly the sector of activity, sales, and numbers of employees. He did not use R&D expenditures or R&D personnel headcount here, because of difficulty to find the data. It is unfortunate that it turns out only to be practical to study R&D for the biggest firms, while for other firms special surveys are needed.

Section 2 of Chap. 8 illustrates issues and difficulties that must be surmounted before carrying out a regression based microeconomic exercise to forecast total filings using a random sample. Frietsch's sampling frame is based on two distinct groups of filings: EPO first filings that were made in 2000 and do not include priority references, and EPO filings that refer to priority forming filings of 2000 from anywhere else. This is a logical way to take account of first patenting activity in the base year. However, our forecasting problem for business planning is to predict total EPO filings without respect to the year of any associated priority filing. Conversion of forecasts from one setting to the other is not hard, but there may be problems caused in econometric models that include autoregressive terms, where an issue arises of whether to use lagged EPO filings, lagged priority forming filings, or both.

We previously looked at recurrent filings behaviour among applicants to the EPO. In many cases the small applicant will file in one year, not in the next, but will file again in some subsequent year. Thus it may be better to construct a transition matrix of probabilities of refiling at different lags to analyse these behaviours. The topic is potentially important for applicant panel survey results, because the projections otherwise ignore the contribution to be made by new applicants who did not file in the base year.

The study of the random sample did not get far enough to actually make forecasts, but we hope that progress can be made in the future because it is important to take account of the entire applicant population and not just the largest enterprises. Frietsch points out that an issue of concern is to improve data availability at the EPO on numbers of patent applications from specified applicant companies. It has long been realised that this is a fundamental problem with patent statistics, especially due to ambiguities in the ways that applicant names are recorded. For legal reasons, EPO records the names and addresses exactly as they appear on the application documents. This can lead to many small variations in the records for a single applicant. We hope that a new client database system that has been installed at the EPO will facilitate the provision of data in-house, particularly using additional company information that gets stored with electronic patent application filings (Epoline 2006). The lack of harmonisation of the client lists that have been available up to now has hampered such studies and may have led to overestimation of the numbers of applicants in the applicant panel survey reports and other places. This is a general problem with patent statistics. A project has been set up by the European Commission to address the issue within the OECD-led Patent Statistics Task Force (OECD 2005).

In Sect. 3, in the study of large companies by combining DTI scoreboard with patent data, the correlation analysis is interesting but practical difficulties are encountered that will need to be addressed before considering implementation. We suppose that there is some kind of natural increase in heterogeneity between R&D and patenting as the years considered become more distant. This will tend to lower correlations and so works against any positive correlation after a long enough lag between R&D and subsequent patenting. Without making corrections for this effect, there could in fact be evidence for a significant lagged effect simply because the correlations do not fall off over time as much as they should.

Dikta has searched for such effects on the macro scale in Chap. 6 and applies transformations to time series before determining the cross-correlations. There is an interesting issue as to whether the same kinds of preparatory transformations that have been proposed by Dikta on macro data can be used at the level of individual companies, or whether there are different requirements for these micro data. An autoregressive model of filings from a single company should usually work, since the operations of an average company change only slowly from year to year. The autoregressive model captures the slow business cycle of the individual company, and comparison with a macro model may perhaps indicate only idiosyncrasies.

The necessity to fit the correlation model backwards against the arrow of time is unfortunate. This arose because the data provided were specified in terms of applicants with priority filings in 2000, while the numbers of EPO filings for these companies were only readily available for the earlier years. Of course, the statistical properties of backwards pointing and forwards pointing regressions are similar. But a problem with the use of forward pointing regressions is data integrity for the most recent years. The incompleteness of euro-PCT-IP filings records for recent history is a major cause for forecasting inaccuracies, although WIPO has been working to improve the availability of timely data (Zhou 2004). Thus, without trying the methods in the forward pointing direction, it is hard to make a fair comparison between the suggested methods and alternative possibilities.

These micro models could be tried again using better data collected over a longer time period. But we wonder whether the approach suggested in Chap. 8 is the most appropriate way to use microeconomic information to augment the macroeconomic regression based methods for forecasting. As with some other methods that have been described in this book, Frietsch's approach can uncover truths about details of patenting behaviours, while our concern is only to forecast numbers of filings. From this latter perspective, the best way to incorporate micro-data is more likely to be by tentative exploration to support the existing macro approaches.

A way forward might be to carry out covariance analysis between individual applicant filing patterns and total filings per sector or country. The applicants with highest covariance (positive or negative) can be selected and these benchmark applicants (or small groups of applicants) will show a patenting history that maps closest to the overall development of the filings total from year to year. We imagine that this might work in a similar way to the capital asset pricing model for company equities on the stock markets (see e.g. Rutterford 1983; Pastor 2002). Perhaps it is too much to ask that the total EPO patent filings can be modelled effectively in this way, but it could be achievable for the totals from particular applicant countries or in certain technical fields or joint clusters. It is not necessarily the case that the best indicator company is one of the largest entities. This kind of study should follow naturally from the macro studies on behaviours of whole countries as were discussed in Sect. 3.6.

A worthy aim of Frietsch's micro-analysis is to examine the causative processes leading to patent application decision making within firms. If communalities can be found, then progress becomes likely for methods that forecast total patent filings via the scrutiny of time series of published company accounts or databases such as the DTI scoreboard. It is clearly advisable to make a comprehensive analysis of these results and underlying databases as well as data that have been collected within the EPO applicant panel surveys over several years. Frietsch has identified many different types of behaviour among companies and it may turn out that few generalisable results can be derived. However, we rather wonder whether success is likely since the effects have not yet become very apparent in other studies.

Frietsch has used a random sampling approach for the selection of companies, but with a sampling scheme that is different to the one that is normally used in the EPO's annual applicant panel survey. In Sect. 2.1.1 of Chap. 8, he favours stratified random sampling rather than simple random sampling of applicants, with strata corresponding to differing classes of numbers of applications per applicant. Various justifications for stratification are given.

For the regular applicant panel surveys that are carried out at the EPO (Sect. 2.3 above), the approach that is taken for the random group is a simple random sampling of applications in the base year. Derivations of formulae for estimators and variances are given in earlier applicant panel survey reports (see 2001 report Annex 3, 2003 report Annex 4, EPO 2006). The distribution of numbers of applications per applicant is highly skewed, in the sense that there are a large number of applicants who make very small numbers of applications (1 or 2) and relatively few applicants who make large numbers of applications. The probability of inclusion in the sample for an applicant is almost proportional to its numbers of applications (but differs slightly because "multiple hits" can occur of the random numbers within the same applicant). The effect of the sampling scheme is to give a fairly flat sample distribution in terms of numbers of applications per applicant. This achieves in practice something similar to the stratified approach that is used in the current study.

Frietsch quotes four reasons, as given by Lee (1989), in favour of a stratified sampling approach. We believe these factors are not crucial in the case of the sampling method that is used for the applicant panel surveys. Firstly (1), there is no evidence that an applicant's intentions towards filing growth rates depend particularly strongly on the numbers of applications that it has previously made, so the strata would not be especially homogeneous. Then (2), there is no special reason for the EPO to wish to estimate growth rates in filings for individual strata that are determined by numbers of applications, since the main purpose is to give a forecast for the growth of the overall number of applications. From the point of view of practicality (3), there is no advantage to administer the field work using strata depending on numbers of applications. Finally (4), there is also no need to use separate sampling schemes or even questionnaires for strata depending on numbers of applications. Optimal allocation – in terms of sample stratum sizes being proportional to population stratum sizes weighted by within stratum variances – is also not relevant in this case because the relatively small number of larger applicant companies account for a big proportion of applications and consequently should be given an increased probability of selection.

Rather than stratifying the sample with respect to numbers of applications, there might in fact be more reason to stratify the samples by bloc of origin. However there are practical reasons why stratified sampling may be inadvisable:- simple random sampling of applications is easier to organise, a simple random sample can be analysed more easily in terms of post-stratification by whatever study variables seem appropriate and there need not be any significant loss of precision compared to a stratified sampling scheme (Cochran 1977, p. 134, Särndel et al. 1992, p. 266).

On other issues of comparison of Frietsch's study to the EPO applicant panel survey:

- The use of a finite population correction for calculation of the standard error of the forecasts is something that could also be considered for the applicant panel survey method and should reduce the apparent widths of the confidence limits.
- Frietsch refers to the fact that large companies such as Siemens are active in several technical areas. Within the domain of the EPO survey, this is dealt with this by asking applicants to define their activities according to the EPO joint cluster definitions and allowing more than one cluster to be indicated.
- In Sect. 2.2, an estimate is given that the proportion of free inventors among applicants to the EPO is almost 28%. We find this figure rather high. In the 2005 applicant survey (EPO 2006), an estimate of 8.3% individual inventors is given. The truth presumably lies somewhere between these margins. Apart from differences in the cohorts of applicants

under study, these results can partly be explained by sampling variation and differential non-response. Frietsch has included applicants for euro-direct and euro-PCT-IP patents, while the survey considers applicants for euro-direct and euro-PCT-RP patents. The latter are normally applied for some one and a half years later than the original euro-PCT-IP application, by which time some individual inventors may have passed on their potential patent rights to larger firms for exploitation. Regarding small and medium sized enterprises (SME), the degree of agreement between the estimated proportion in Sect. 2.2 and that in the 2005 applicant survey is better – Frietsch estimates 65% while the survey estimates 71%.

- In Sect. 3.2, an estimate of the average R&D expenditure by R&D intensive firms in 2000 is given as £392 million. A recent estimate of average R&D spend per firm in 2004 was obtained from the 2005 applicant survey as EUR 1.2 million per patent filing (EPO 2006). A weighted average number of first patent filings per respondent from the survey is 111, making an average R&D spend per firm of about EUR 133 million. Reasons for the discrepancy between £392 million and EUR 133 million may include the fact that the survey estimate includes somewhat smaller applicant firms, even though it is still skewed towards larger entities. Statistical treatment to compensate for skewness in the distributions of firm sizes requires further research.

- In Sect. 3.3, Frietsch finds that the highest correlation between patenting and R&D is in the contemporaneous year. This apparently conflicts with the survey result of an average 14 month lag (EPO 2005), as well as with other even longer estimates that have been found from macroeconomic regressions in this volume. A possible explanation for this is the argument given earlier that there can be a confounding of company growth effects in both variables with causative factors in the correlation matrix.

4 Implementation of suggestions at EPO

The work of the research programme has given several suggestions for the development of forecasting procedures at the EPO. There are also indications for further research. We have the modest immediate requirement to incorporate at least the most practical and useful methods directly into our annual forecasting process. For this purpose, initial attempts were made to utilise versions of Dikta's bloc based approach and Park's countries based approach in the annual forecasting exercise of January 2005.

A version of Dikta's ADL approach (Chap. 6, Sect. 5.3) was applied to data on patent filings from 1979 to 2003. No distinction was made between types or modes of applications (euro-PCT-IP vs euro-direct; first vs subsequent) or technologies. The R&D data that were used were based on data of Business Enterprise expenditure on R&D excluding government expenditures (BERD financed by industry), expressed in terms of US $ as of year 2000 and obtained from the OECD MSTI series. The five year lag between R&D investment and EPO patent filings that was established in Chap. 6 was confirmed. The resulting forecasts were better than the self-determined forecasts, because the calculated standard error of the filings forecast for 2004 was slightly smaller in each bloc for the model involving R&D as predictor than for the self-determining model. The model could provide forecasts only out as far as out to 2006, which reflects the fact that complete R&D data were available only up to 2001, and this maps to 2006 with the suggested five year lag. The results were rather disappointing in the sense that the forecast for total filings in 2004 was considerably less than the actual observed figure. We intend to move forward with this method using the S-Plus package (Insightful 2001) and by making full use of Dikta's conclusions.

A variant of the methodology of Park (Chap. 7) was also used in early 2005. Park implemented this model in STATA, but we have developed it using SAS (SAS 2006) and Excel (Microsoft 2006). The MODEL procedure of SAS is useful to perform the multi parameter fitting exercise, while Excel is straightforward to use for data manipulation and visualisation of the results. Historical data on EPO filings were considered annually, after partitioning into 30 most important source countries and one other heterogeneous residual area (ZZ) to capture the remaining countries. No distinction was made between types of applications or technologies. Data were standardised between countries by dividing by the size of the labour force and taking natural logarithms.

For each country of origin, the series of annual EPO patent filings was used together with series of the annual size of the labour workforce and a series of annual national R&D expenditures, based on R&D data that were obtained from BERD financed by industry, again as obtained from the OECD MSTI series. This time the BERD data were deflated by purchasing power parities (PPP) and expressed in terms of US $ as of year 1995.

In the case of the residual area ZZ, the labour and R&D figures were obtained by amalgamating figures for all reporting OECD countries that fall outside the 30 specified countries, while the patent counts comprise all EPO filings that are not from the 30 countries.

The approach involved a multiple regression model with 5 parameters: an intercept, three autoregressive terms for the past 3 years filing totals per country, and 5 year lagged R&D per country, after taking the same standardisation transformation on the independent variable (R&D) that is used for the dependent variable (filings). The five year lag was chosen because of Dikta's ARIMA based transfer function result (Sect. 3.5).

The model can be written as follows.

$$Y_{t,j} = a + b_1 \cdot Y_{t-1,j} + b_2 \cdot Y_{t-2,j} + b_3 \cdot Y_{t-3,j} + c \cdot X_{t-5,j} + error_{t,j}$$

Where

$$Y_{t,j} = \log \left[\frac{\text{Number of filings in country j in year t}}{\text{Number of Workers in country j in year t}} \right]$$

$$X_{t,j} = \log \left[\frac{\text{Discounted R\&D in country j in year t}}{\text{Number of Workers in country j in year t}} \right]$$

$j = 1, ..., 31$; $t = 1980, ..., 2000$; or $t = 1980, ..., 2003$.

This model was fitted simultaneously, without allowing for parameter variations between countries. The expected values for filings (and forecasts) were then recovered by reversing the standardising transformation.

Annualised data on patent filings per country from 1980 to 2003 were analysed. In the first experiment, the model was fitted on the subset of data up to year 2000. Table 9.1 shows the parameter estimates from this model and gives forecasts of total filings for years from 1986 to 2008. After 2005, there are no longer lagged R&D data available for all 31 countries and filings forecasts can be made only for a subset. These subset forecasts have then been scaled up, using division by a weighting factor that corresponds to the share of 2003 filings given by the subset. The plots in Fig. 9.2 show actual and fitted values for total EPO filings from the whole world and also EPO filings from residents of Belgium, Germany, France, Great Britain, Japan and US. Visual inspection suggests that the best indicator country for total filings is Belgium[2].

This model gives a bad fit in terms of total filings to year 2001, but the deviation for 2002 is not big and there is an excellent fit for 2003 and 2004. It does not follow that the forecasts for each country are necessarily so good, which is probably due to the insistence to fit a small number of parameters with common values for each country.

---

[2] Belgium is also one of the countries for which Blind found a positive relationship between R&D expenditures and patenting in Chap. 5.

Table 9.1. EPO forecasting model based on Park's methodology. Forecasted filings and estimated model parameters

Year	Actual Filings	Forecasts for model fitted to years	
		1980 to 2000	1980 to 2003
1980	20 012		
1981	25 487		
1982	28 955		
1983	32 145		
1984	37 507		
1985	39 988		
1986	44 096	44 508	44 375
1987	48 343	49 110	48 977
1988	55 894	53 734	53 592
1989	62 597	60 967	60 865
1990	70 955	68 699	68 557
1991	67 584	77 123	76 980
1992	70 345	77 786	77 395
1993	70 791	79 539	79 198
1994	74 250	80 658	80 288
1995	79 261	83 715	83 378
1996	87 405	88 381	88 073
1997	100 392	95 836	95 586
1998	113 342	107 428	107 274
1999	123 878	120 722	120 569
2000	145 241	132 885	132 641
2001	162 020	143 100	151 300
2002	161 068	153 649	169 766
2003	167 205	164 606	174 841
2004	178 843	176 710	186 237
2005		189 283	198 508
2006		204 032	211 269
2007		216 414	222 790
2008		243 747	242 039

	Estimate (S.E.)	Estimate (S.E.)
Intercept	-0.885	-0.849
	(0.276)	(0.236)
AR 1	0.637	0.657
	(0.05)	(0.041)
AR 2	0.251	0.237
	(0.051)	(0.048)
AR 3	0.046	0.041
	(0.040)	(0.037)
R&D	0.063	0.060
	(0.023)	(0.020)

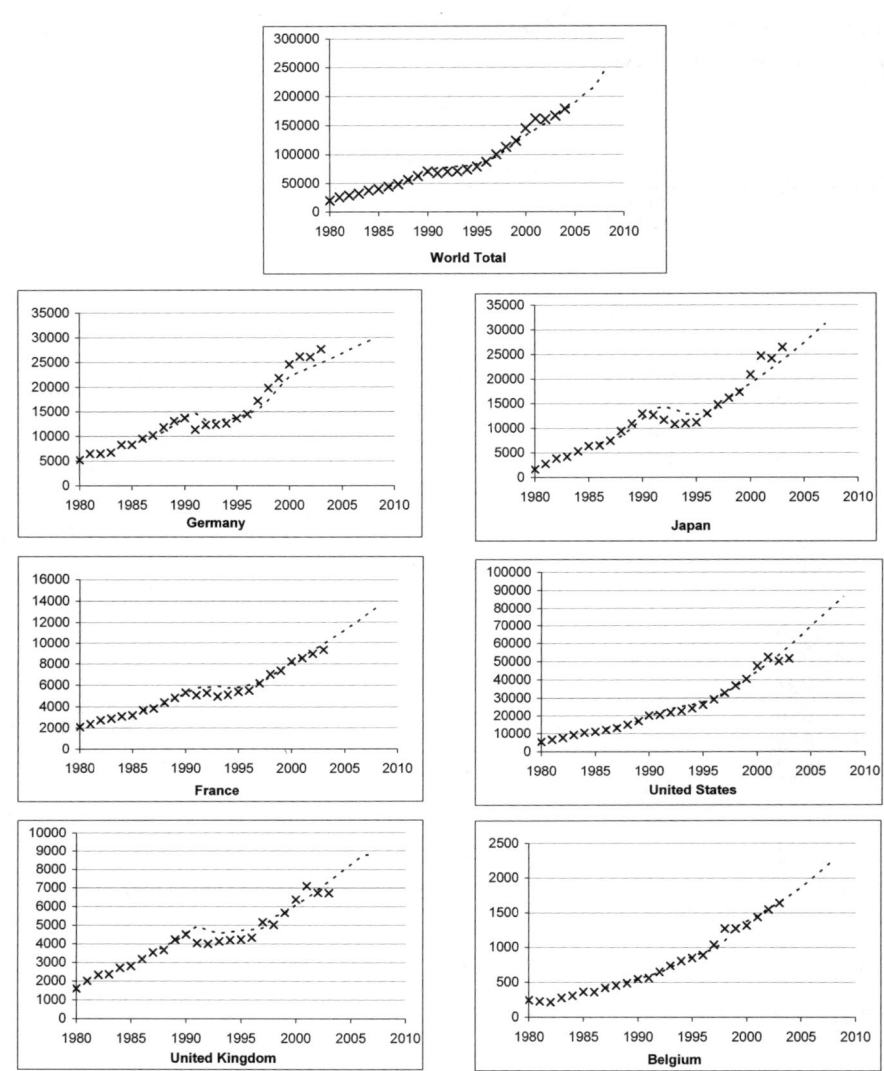

Fig. 9.2. EPO forecasting model based on Park's methodology. Actual (x) and forecasted (---) filings at EPO from residents of selected countries from the model fitted to years from 1980 to 2000

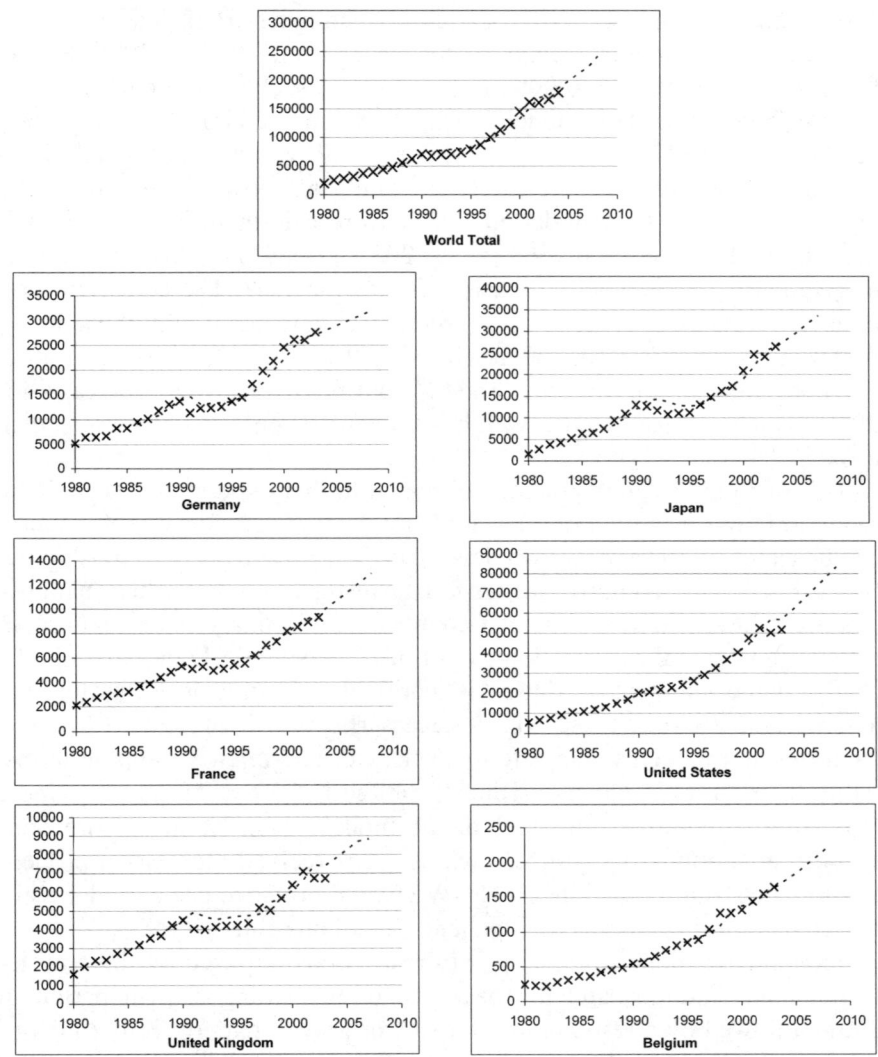

Fig. 9.3. EPO forecasting model based on Park's methodology. Actual (x) and forecasted (---) filings at EPO from residents of selected countries from the model fitted to years from 1980 to 2003

In the second experiment, the model was fitted to the data from 1980 up to 2003. Table 9.1 also shows the parameter estimates from this model and gives forecasts for total filings in the years from 1986 to 2008. Subset forecasts for years 2006 to 2008 have also been scaled up in the same way as described above. This model does not give such a good fit as the previous one to filings in 2003 and 2004, but is still within an acceptable forecast error range of 5%. The forecasts from 2005 onwards are somewhat higher than in the first model. One way to rescale them is to standardise them by multiplying by 0.9563 (= 167 205 / 174 891), so that they appear to give perfect agreement with actual filings for 2003. The corrected forecasts for years 2005 to 2008 come into closer agreement with those in the first model (2004: 178 103, 2005: 189 838, 2006: 202 042, 2007: 213 060). For both models, not too much trust should be placed in the forecasts for 2008, which are based on a weighting factor of only slightly more than 60%.

This method has good potential although further experiments seem to be required to put it into an optimal form for routine use. Since each country can be expected to have its own developmental pattern and relationship between R&D and patenting, we wish to investigate models where parameters are allowed to vary between countries. Preliminary investigations in SAS suggest that the regression slopes of patents vs R&D do vary significantly, while intercepts variation is marginally significant. Regarding the autoregressive terms, these do not necessarily vary significantly between countries and it seems adequate to work with one or two common terms, rather than three. In Chap. 5, Blind explored the effect of an extra linear trend term (trending with time) in the models in order to capture non-specific institutional variations, such as the perceived increase in propensity to patent during the late 1990s. We would prefer to use a model without such a linear time trend component if at all possible.

Econometric models are subject to restrictions imposed by the real developmental patterns, and the peak out of the strong growth period in 2001, followed by a resumption of a slower growth pattern from 2003 onwards, is going to challenge any model. This is probably the reason why the forecasts for 2003 and 2004 filings are better on the model fitted up to year 2000 than on the model fitted up to year 2003. With the benefit of hindsight, it can be seen that the model of Fig. 9.3, that used data up to 2003, tended to overestimate filings for most countries in 2001 and 2002, when there was a retrenchment after the exuberance of the late 1990s.

The distillation of appropriate R&D series as input variables for the models is an important consideration. In our attempt to use the Dikta and Park approaches, we tried aggregated BERD financed by industry. But Park prefers to use gross R&D expenditure (GERD), and used BERD only for technology group based sectoral analyses. He finds it hard to believe that government-financed BERD has no effect on patenting in the light of what is known about spillover benefits and technology transfers between government and commercial laboratories. Knowledge from the public domain or contributed by the public sector has often influenced private sector technological development (e.g. world wide web, medicines, etc.). These days, private firms can often apply for patents on technological developments that were funded by government. Perhaps therefore, industry financed BERD could be used for the early part of the period covered by the data and then the rest of GERD can be phased in during the 1990s, from a point where it can be believed that it led to patentable inventions. An understanding of distributions and trends in global R&D statistics clearly would be useful in order to apply this model properly. There is as much need for up-to-date and accurate data on R&D as there is for the patent series themselves.

Both the approaches that we have tried so far show promise, but only Park's approach delivered forecasts that proved to be useful for the early 2005 budgeting exercise to produce a six year plan. However, the comparison has been unfair in the sense that no check was made with the Dikta method fitted to data up to year 2000 only. Perhaps the now-casts for 2003 and 2004 would then be just as good as under Park's model.

In the course of preparation for the January 2006 forecasting exercise, versions of the Dikta and Park approaches were tried out again as well as the existing approaches. Park's approach provided optimistic forecasts that lie above the results of the traditional methods, while the forecasts from Dikta's approach came out below those of the traditional methods. It is remarkable, that two methods that purport to be based on a consideration of similar economic factors give such differing results. There is clearly more work to do in order to find a synthesis between these methods. An interesting avenue for experimentation will be to try out Dikta's data preparation techniques, such as Box-Cox transformations, differencing and pre-whitening, on the input data to Park's econometric formulations. Another possibility is to apply Dikta's approach on individual countries, in the way that Park did.

Regarding the approaches that are reported in the other chapters, we have learned that the analysis of monthly data may sometimes be appropriate even when the goal is to make annual forecasts. We see usefulness in Meade's methodological approach and software recommendations. We intend to explore the scope of application of state space modelling as a complement to Dikta's transfer function and multivariate approaches. We do not presently intend to pursue Blind's approach, unless we are considering forecasts to be made at more detailed levels of geographical regions or technical areas, an approach that Meade has suggested may not be required to give more accurate forecasts for total filings. For the future development of microeconomic based forecasting approaches, extensions and enhancements of Frietsch's work as well as of the applicant panel survey will be considered. We encourage others also to try these approaches, either on EPO patent data or on their own systems.

5 Comparisons of forecasts

In Sect. 4.1 of Chap. 2, Harhoff mentions the problems of selecting or combining forecasts that have been made by different methods. Every candidate model should be evaluated on data sets that span different groups of years, in order to cope with heterogeneity in the data generating process and with effects that hit the process from outside the modelled variables. It could well be that each phase of the economic cycle produces a different model that works best. If this can be determined systematically, then a portfolio of models can be applied with confidence.

We could attempt to compare the methods that have been suggested by the authors of Chaps. 3–8 via an assessment of forecasting accuracy when applied retrospectively over the patent filings series that were available in successive years. Meade demonstrates this approach in the discussion around Fig. 4.5 in Chap. 4. Statistics which can be calculated for the assessment of forecasting ability of different methods include RMSE and MAPE (Chap. 4), and RMSPE (Chap. 7). These statistics should be constructed over various horizons from differing starting points.

However, in the context of the annual forecasting procedure to develop scenario forecasts at the EPO, each method produces several alternative forecasts by variations in the modelling assumptions, which makes it difficult to apply such a retrospective historical assessment. Also, the methods will have different behavioural characteristics in different situations. All regression based methods are likely to cheerfully predict further increases after a period of steady growth. On the other hand, the econometric based methods might be expected to be in a better position to signal possible turning points to a pre-existing trend. Thus, an ability to predict turning points or to contribute to *what-if* analyses is an alternative and complementary feature to that of raw forecasting ability.

In practice, a suite of forecasting approaches will probably be retained instead of concentrating on just one or two methods. But ways to compare results in a fair way should be developed.

Consider methods to provide either a direct choice of a favoured scenario amongst a group of alternative choices, or to make an appropriate weighted average scenario, giving as weight to each scenario a measure of relative confidence in its message. Within the scope of methods that have been suggested for dealing with alternative non-nested models in the statistical literature, attractive techniques that can achieve these aims are Bayesian model discrimination (BMD) and Bayesian model averaging (BMA). These methods are closely related and the term *Bayesian* indicates that a prior probability of the correctness of each candidate model can be given. The evidence of goodness of fit of each model to the past data can be used to modify these prior beliefs to give posterior weights for each model. (Davison 2003; Zhao et al. 2005; Hingley 2006).

Within each method (trend, transfer, econometric), the BMA is developed as follows, where p( ) indicates a probability density function and d candidate models are indexed by i, where $1 \leq i \leq d$.

A probability density function is developed for the forecast conditional on the data by averaging over models.

$$P(\text{forecast} \mid \text{data}) = \sum_{i=1}^{d} p(\text{forecast} \times \text{model}_i \mid \text{data})$$

$$= \sum_{i=1}^{d} p(\text{model}_i \mid \text{data}) \cdot p(\text{forecast} \mid \text{data} \times \text{model}_i)$$

A model weight is given by $w_i = k \times p(\text{model}_i \mid \text{data})$, where k is an unknown constant. Then the above equation becomes

$$P(\text{forecast} \mid \text{data}) = \frac{\sum_{i=1}^{d} w_i \cdot p(\text{forecast} \mid \text{data} \times \text{model}_i)}{\sum_{i=1}^{d} w_i}$$

The BMA forecast is obtained by taking the expected value of this density, which is equivalent to the weighted average of the expected values from each scenario.

$$\text{BMA forecast} = \frac{\sum_{i=1}^{d} w_i \cdot E(\text{forecast} \mid \text{data} \times \text{model}_i)}{\sum_{i=1}^{d} w_i}$$

In the 2006 forecasting exercise, a pragmatic initial approach to calculate a BMA was taken. Rather than developing a relative goodness of fit measure from the likelihood function of each model, the estimated residual standard error from past data was used to give the posterior weights. It is easier to fit a model to a shorter series of data values, so shorter series were positively discriminated against by multiplying the estimated standard error by the two sided 95% critical t value with the same number of degrees of freedom[3]. The reciprocal of the measure was then taken as the weight for each scenario. This pragmatic choice for the weights w_i is

$$w_i = \frac{1}{t^*_{vi} \times s_i}$$

Where t^*_{vi} is a 95% two sided critical t value on vi degrees of freedom, and the residual standard error over n data points is the square root of the residual variance s_i^2.

$$s_i^2 = \sum_{j=1}^{n} \frac{(y_j - y_{jpred})^2}{vi\,(n-1)}$$

The BMA conditional forecasts, conditional on the observed historical data, were calculated separately for the results that were obtained from the trend methods, the transfer methods and the econometric methods. A simple average of these three forecasts was then made to give an overall forecast scenario. The BMA scenario generated in this initial exercise for the early 2006 forecast exercise was almost identical to a linear regression fitted to total EPO filings since 1996 but excluding the years 2000 and 2001 from the fit (see Chap. 1, Fig. 1.1). Thus, in a sense, the methods come full circle to confirm the pragmatic initial analysis. Further investigation is

---

[3] Degrees of freedom = Number of data values − Number of estimated parameters.

clearly needed before confirming such a simplistic conclusion on this issue.

For Bayesian model discrimination (BMD), the candidate model with the highest w_i value is selected. BMD would seem to be a technique of choice only where a forensic examination of the underlying mechanisms is the goal. This does not seem adequate for the EPO forecasting problem, where there is reasonable doubt that any particular model is exactly correct. Any errors of formulation in regression based models for macroeconometric analysis can lead to forecast bias, even if problems are only in the minutiae. Formal robustness testing for deviations of the estimation model from assumptions about the data generating model is a desirable goal, which might lead to conditional acceptance of a portmanteau model, even in the knowledge that it differs from the data generating model (see e.g. Swamy and Tavlas 2005; Hingley 2004, 2006).

6 The future of forecasting at the EPO

The approaches that have been reviewed are mostly annual forecasting methods, but there is scope to move towards making more frequent forecasts on a seasonal or monthly basis. This applies especially for the joint cluster departments of the EPO, where fluctuations of the business plan can have a more destructive effect on operations than are experienced at the centre. Exponential smoothing, and other related methods that can be incorporated in a state space approach, seem to be good candidates for this (see Sect. 3.3).

Within the domain of the conventional regression approach, several other techniques are available. These include non-parametric methods (Gibbons and Chakraborti 2003), robust regression methods (Wilcox 2005) and functional data analysis (Ramsay and Silverman 1997). We can not claim that this list is exhaustive and other possibilities may emerge.

An alternative technique to regression based approaches and surveys of applicants is the forecasting competition. Rather than a research competition of the kind suggested by Harhoff (Sect. 3.1 of Chap. 2), this is an exercise in which interested participants make quantitative forecasts by whatever methods they wish and then the one who comes nearest to the out-turn wins a prize. Interested parties could be those who have special knowledge of the patent process, including EPO examiners, patent attorneys and inventors. A disadvantage of this technique is that it may be more useful as an incentive rather than as a planning method, unless the same winning players turn out to have good and reproducible forecasting ability over a number of years.

In addition to constructing forecasting machinery along the lines that have been suggested by the work of Dikta, Park and the other authors, we would like to consider a generalisation of the transfer approach of Sect. 2.2. As Harhoff mentions in Sect. 3.2.3 of Chap. 2, global models can take account of econometric factors and first filing/subsequent filing patterns throughout the world. No patent office should consider its forecasting problem in isolation. Just as the applicant will decide which of the various the world patent offices to use, so should each office consider all world wide inventions as its potential market.

The patent families approach remains the one that is most likely to lead to improved econometric models for understanding the structure of the world wide patenting system. With patent families, it may be possible to model patenting flows efficiently and so provide simultaneous forecasts for filings in all major patent systems. The potential for this has led to interest within the trilateral statistical working group. Its work in this direction has not yet progressed very far however. In the earlier chapters, Dikta has mentioned the analysis of flows in Chap. 6, but only Park has considered this aspect in depth in Chap. 7, by separately forecasting EPO first filings, EPO subsequent filings, and patent families. Unfortunately, as mentioned in Sect. 3.6 above, the available patent family data are calculated from publications databases and so are never completely up to date. It is important to try to augment these data sets as far as possible with latest live filings data, in order to allow them to be used properly for forecasting purposes.

A simultaneous equations based system for international filings flows may be useful for global modelling from patent families data. The patenting flows between blocs depend on domestic filings in each bloc, the attractiveness of international markets and the extent of globalisation of production and consumption. The processes of fitting models to data involve problems of identification for the various parameters. For estimation purposes, it may be simpler to incorporate lagged terms for the independent variables and/or the dependent variable itself. Seemingly unrelated regression may be an appropriate estimation technique (Zellner 1962; Maddala 2001)[4].

But it is also interesting to see what happens when the lagged terms are ignored and consideration is restricted to a system of contemporaneous modelling terms. There may have been too much concentration within the

---

[4] This technique was already mentioned in the time series context as SUTSE in Chap. 4, Sect. 3.

research programme on technicalities of regression modelling at the expense of structural descriptive modelling of the system under study. After all, we are working with known counts of the total numbers of patent filings and therefore already have a complete data description of the system that is under study.

As an example of this approach, consider a simple model for subsequent filings that arrive at each of the major trilateral offices. This is an extension of the transfer approach of Sect. 2.2 and is applied to filings data rather than to patent families data, even though it uses the priority forming first filings information in the regions of each of the offices. Countries outside the trilateral region are ignored in this example. The filings flowing from one bloc to another (F) are considered to depend on domestic filings (D) in the origin bloc, on competitive factors represented by the contra-flow of filings from the target bloc to the origin bloc, and also on the flow from the third bloc to the target bloc (Fig. 9.4).

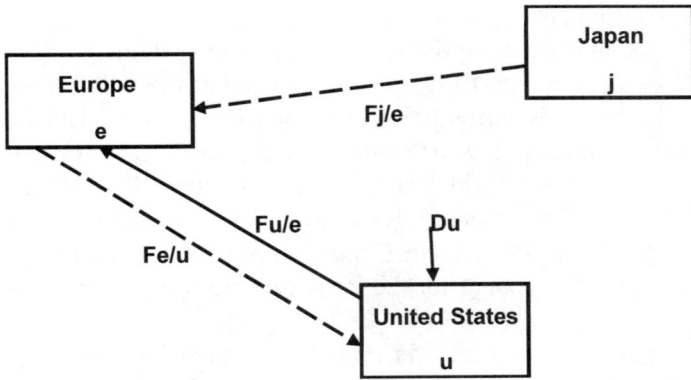

Fig. 9.4 Simultaneous equations based model for international patent filings flows, that allows contemporaneous contributions from other flows. This diagram shows the components of the first equation below for $F_{u/e}$

$$F_{u/e} = b_1.D_u + e_1.F_{e/u} + f_1.F_{j/e}$$
$$F_{u/j} = b_2.D_u + e_2.F_{j/u} + f_2.F_{e/j}$$
$$F_{j/u} = b_3.D_j + e_3.F_{u/j} + f_3.F_{e/u}$$
$$F_{j/e} = b_4.D_j + e_4.F_{e/j} + f_4.F_{u/e}$$
$$F_{e/u} = b_5.D_e + e_5.F_{u/e} + f_5.F_{j/u}$$
$$F_{e/j} = b_6.D_e + e_6.F_{j/e} + f_6.F_{u/j}$$

Where D_e, D_j, D_u indicate numbers of domestic patent filings in Europe, Japan and US respectively;

$F_{u/e}$, $F_{u/j}$, ..., $F_{e/j}$ indicate the flows of subsequent filings from US to Europe, from US to Japan, ... etc. ..., from Europe to Japan.

In this system, it can be seen that the left hand side terms of each equation also appear on the right hand side of two of the other equations. The set of simultaneous equations can be solved, either by hand or with the help of computer algebra software such as Scientific Notebook (MacKichan Software 2006), but the solution is a complicated mixture of terms that is not easy to specify in a regression model. Further, there are dependencies between coefficients in the model that lead to constraints on the values that can be taken.

How far can we go with a description of this system while avoiding complicating the analysis with regressions? Inspection of the equations suggests that the domestic filings terms $b_1.D_u$ etc. are essentially equivalent to generalisations of the transfer model, while the competitive flow terms $e_1.F_{e/u}$ etc. are new components.

The data for D_e, D_j, and D_u comprise, as far as possible, domestic first filings in the EPC area (including EPO), at the JPO and at the USPTO respectively[5]. Then the six flow terms are represented by reported data on subsequent euro-direct filings (EPO) or national filings (JPO, USPTO) appearing from the other blocs, to which are added subsequent PCT-IP filings from the other blocs that involved designations of each office[6]. Data were analysed by year of the subsequent filings, with the domestic filings series advanced by a year to account for the Paris Convention priority year effect.

In a first step, a global six way transfer modelling scheme can be introduced by setting the e and f terms to 0. Table 9.2 shows the results of a trial analysis of this sort.

---

[5] For USPTO only, due to lack of information on breakdowns of the data, D_u contains national domestic first and subsequent filings, as well as domestic PCT national phase entries.

[6] For USPTO only, due to lack of information on breakdowns of the data, the flows $F_{j/u}$ and $F_{e/u}$ included national first and subsequent filings from the other blocs, as well as PCT national phase entries from the other blocs.

Table 9.2. Simultaneous equations based model for trilateral patent filings flows. Transfer model assuming no competitive flows

Flow equation	1	2	3	4	5	6
Flow	$F_{u/e}$	$F_{u/j}$	$F_{j/u}$	$F_{j/e}$	$F_{e/u}$	$F_{e/j}$
e_i fixed for equation i	0	0	0	0	0	0
f_i fixed for equation i	0	0	0	0	0	0
Year	b_1	b_2	b_3	b_4	b_5	b_6
1994	0.228	0.223	0.125	0.034	0.415	0.199
1995	0.231	0.222	0.138	0.041	0.450	0.216
1996	0.222	0.214	0.134	0.044	0.452	0.232
1997	0.291	0.278	0.143	0.048	0.486	0.252
1998	0.288	0.276	0.152	0.049	0.541	0.271
1999	0.282	0.266	0.159	0.058	0.601	0.285
2000	0.300	0.286	0.180	0.070	0.628	0.310
2001	0.303	0.287	0.197	0.063	0.670	0.329
2002	0.270	0.255	0.197	0.069	0.693	0.338
2003	0.261	0.245	0.218	0.087	0.692	0.342
Regression slope	0.006	0.005	0.010	0.005	0.035	0.017
R^2 statistic for regression	0.35	0.28	0.95	0.92	0.97	0.99

Table 9.3. Simultaneous equations based model for trilateral patent filings flows. Includes assumed values for competitive flow effects

Flow equation	1	2	3	4	5	6
Flow	$F_{u/e}$	$F_{u/j}$	$F_{j/u}$	$F_{j/e}$	$F_{e/u}$	$F_{e/j}$
e_i fixed for equation i	0.3	0.05	0.2	0.3	0.5	0.5
f_i fixed for equation i	0.1	0.1	0.2	0.1	0.4	0.2
Year	b1	b2	b3	b4	b5	b6
1994	0.089	0.182	0.084	0.007	0.150	0.104
1995	0.087	0.181	0.092	0.010	0.170	0.111
1996	0.096	0.177	0.088	0.013	0.158	0.115
1997	0.128	0.231	0.091	0.012	0.175	0.127
1998	0.123	0.230	0.095	0.011	0.208	0.139
1999	0.111	0.221	0.096	0.017	0.255	0.141
2000	0.129	0.240	0.109	0.022	0.250	0.146
2001	0.131	0.240	0.123	0.014	0.258	0.168
2002	0.106	0.211	0.122	0.019	0.287	0.172
2003	0.102	0.201	0.140	0.035	0.268	0.153
Regression slope	0.002	0.004	0.006	0.002	0.016	0.007
R^2 statistic for regression	0.20	0.28	0.86	0.65	0.90	0.88

At the bottom of each column of Table 9.2 appear the b regression slope coefficients with their associated R^2 statistics for linear trends over the years in question. For $F_{j/u}$, $F_{j/e}$, $F_{e/u}$ and $F_{e/j}$, the b slope values are significantly positive and indicate an increasing propensity over time for domestic first filings to flow abroad in the following year. For the two flows out of US, $F_{u/e}$ and $F_{u/j}$, the slopes are not significantly positive. This seems to have been associated with a "hump" in the data at the end of the 1990s, as has already been mentioned in Sect. 2.2[7].

Table 9.3 shows the estimated b coefficients after the e and f coefficients have been set at values that correspond to an arbitrary first approximation of the competitive effects. The e and f coefficients for each flow are assumed stable over time. The b values have then been calculated to ensure that the actual flows, that are the subjects of the set of equations, still correspond to the given data values. For the full model, the R^2 statistics for the time trend of each column in Table 9.3 indicate that the regression slopes for trends of the estimated b coefficients over time are somewhat less significant than in Table 9.2. This gives an indication as to how we might be able to approach the question of unravelling the effects on patenting of the inventive process from the effects of increased global competition. It might address some of Blind's concerns regarding disruptions to the relationships between R&D and patenting in recent years. Other modelling schemes are of course possible.

A large organisation, such as EPO, requires long-term indications of future developments in order to give a sense of direction for planning and to ensure financial stability, particularly in areas such as risk assessment of future pension arrangements. An optimal plan might range up to 25 years into the future. The statistical approach to forecasting by regression methods that has been detailed in this book seem rather inappropriate for this task, since confidence limits on forecasts widen beyond realistic bounds within just a few years. Scenario building by groups of experts has a place in these deliberations, perhaps using deliberately pessimistic, realistic and optimistic viewpoints. The Delphi technique is of this type (Armstrong 2001, p. 125). In advanced approaches, scenario building combines quanti-

---

[7] In theory, the b values in the row of Table 9.2 that corresponds to year 2003 could be calculated using the data on domestic filings and flows from Fig. 1.2 of Chap. 1. However, the data sources are slightly different and only the flows $F_{u/j}$ and $F_{e/j}$ are truly comparable, since Table 9.2 represents flows to Europe in terms of filings at EPO and not in other EPC offices, and the counts of flows to the USPTO have been increased somewhat for reasons mentioned earlier in footnotes 5 and 6.

tative data and qualitative assessments systematically (Löfstedt 2003). In fact a scenario workshop project took place at the EPO in 2005.

An important consideration for long term forecasting is whether financial equilibrium is proposed in the distant future. If it is, then it can be assumed that levels of finance and business activity will achieve a balance, even if it can not be said *a priori* what quantity of business activity corresponds to the balance point. Risk assessment is required in addition however. On the other hand, if a lack of balance is anticipated, as in the late 1990s when applications outstripped the EPO's ability to process them, then limits on the degree of divergence have to be set to ensure that a good enough service is given to clients. It seems unlikely that future percentage growth rates will ever exceed the maxima that were experienced in the late 1990s period, and so upper limits to future growth in total EPO filings can be set at a level of about 15% growth per annum. The lower limit to growth remains -100% however, since the stability of the current system against political change can not be absolutely guaranteed.

7 Conclusions

In Chap. 2, Harhoff proposed a typology of research modules for the improvement of forecasting methods for patent filings. After monitoring the results of the programme and relating them to the EPO business activities, we can now propose an alternative but complementary typology. Firstly, there are the pragmatic methods, such as trend analysis and the transfer model. These take account of essential factors, but ignore some details of structure as they search for appropriate forecasts. Secondly, there are the more intensive approaches, that still concentrate on the European patent system alone, but take account of appropriate micro- and macro- factors. Most of the contributions to the research programme have operated at this level. Finally, there is the prospect for integrative global analysis of the world patenting system, where the factors that influence EPO patents will fall out of a mix that can simultaneously forecast activities in all the major world patent systems. The methods usually involve regression techniques, but these can sometimes be avoided by means of surveys, systems theory or judgemental approaches.

So, various methods can be considered for approaching the forecasting problem, and we look forward to making progress with these in the future. However two major problems remain to face forecasters who use almost any method to model data on the volumes of patent filings. These problems are the ever changing nature of the patenting environment and the lack of availability of data on the most recent developments.

Regarding the changing nature of the criteria used by patenting decision makers, this difficulty can be partially resolved by adaptive estimation techniques. But it should not necessarily be assumed that the European patent system will remain as it has been in the past. Over the short history of the system, there have already been changes due to centralisation of patent applications as the EPO established a reputation in the 1980s, increasing international interest with the popularisation of the PCT system in the 1990s, and benefits from a continued lowering of fees in real terms as well as sociological mood change involving IP strategy by companies in the late 1990s and beyond. In the future, political shifts could move in favour of a single patent examination system for Europe and even perhaps for the world[8]. Increasing centralisation of patent systems might benefit the EPO via recognition of its expertise. But for planning purposes, it may need to be understood that the object to be forecasted becomes something different to what existed before. Consideration should always be given to such possibilities before carrying out trend based projection methods of historical EPO filings data sets.

Several specific potential events in a medium term perspective may have significant impacts on the applications at the EPO. In addition to world events, that may or may not be broadly predictable, relevant patent related events include fee policy changes (discussed in Sect. 3.6), as well as several other possible factors:

- The reduction of the requirements for the translation of granted EPO patents in order to be valid at the National Offices of destination. Such a change would imply a substantial reduction of the costs of the European patent procedure.
- Introduction of the European Community Patent (EPO 2003). The impact of this event largely depends on the way that it may be formulated.
- Introduction of the European Utility Model (European Commission 2002).
- Changes in the scope of patent protection, in areas such as business methods which are currently not patentable under the European patent Convention.

---

[8] However, the administrative council of the EPO largely consists of representatives of the national patent offices of the EPC contracting states, and some influences exist there that may favour enhancement of activities of the national patent offices, rather than centralisation (Abbott 2006).

- Changes in the list of countries in which European patents can be validated.

The occurrence or continuation of such factors in the future needs to be taken into account when making forecasts. Such prospective analyses can also encompass *what-if* approaches, in order to provide more rational assessments of the impacts of management decisions or external political developments on the EPO. The development of patenting at the EPO can also be considered in terms of the scenario of the past. It seems likely that the observed development has been caused by a combination of favourable winds, that underlie the increase in the numbers of patent filings since the EPO was instituted. These favourable winds include the factors that were already mentioned above, as well as the increased fashionableness of patents and the globalisation of markets. The continuation of such trends in the future depends on a continuation of an environment of such favourable factors.

The second major problem is that delays are experienced in obtaining patent filing data from some sources. This hinders efficient forecasting, though efforts can be made to circumvent it. Almost 70% of the incoming patent filings at the EPO are euro-PCT-IP filings. Counting euro-PCT-IP filings is difficult because of procedural delays in transferring incoming documents from the many PCT receiving offices around the world to the International Bureau (IB) of WIPO in Geneva. Traditionally, counts have simply been made of the numbers of PCT applications by the dates at which they are received at the IB. While this count has the advantage of being fixed early and then remaining constant, it is often a biased proxy for counts of the numbers of PCT applications by the dates they are filed at the various receiving offices themselves, because of procedural delays. The forecasting group at EPO has often been faced with highly uncertain estimates of euro-PCT-IP filings for the preceding two calendar years. However WIPO has recently developed a now-casting method that should be a considerable improvement.

Other problems exist for analysis of the available patent data however. Difficulties with first filings data from USPTO have been mentioned in Sect. 2.2. It has only recently become the practice in USA to publish applications at 18 months after priority forming first filing, as is done almost everywhere else. Even now, applications are not necessarily published if the applicant declares no intention to make a foreign filing that is based on the USPTO application.

There is also a question of standardising definitions of patent filings to ensure that all interested parties are working with comparable data sets. For this programme, good standardisation was obtained by providing the researchers with similarly defined data from the EPO's in-house systems. For similar studies in the future, a suitable source of comparable data could be provided, for example, by the reference world-wide data set on published patent filings that is under development at EPO for use by accredited researchers of the patent statistics task force (OECD 2005).

The different forecasting approaches that have been discussed aim at short term and medium term predictions of the patent applications at the EPO, out to no further than a six year horizon. They are based on a variety of different determinants for the development of future applications, such as R&D expenditures or changes of applicants' behaviour patterns. All these factors usually have the feature in common that they change in a steady, continuous way. As long as this assumption remains valid, forecasting can progress without necessarily having the most up to date information at our fingertips.

In this book, there has been a tendency to consider forecasting as an isolated problem. However, many people are working on the development of the patent system from complementary angles, such as the legal and economic aspects. A large information supply is available and nobody is an expert on everything. To misuse data mining parlance, we may assert that everyone has only a view of their own data mart within a large encompassing data warehouse (Giudici 2003). All can benefit from synergies and communication between the interested parties is advisable. Just as forecasters need to know about possible changes to the scope of patents, so do administrators need to know about likely future patent demand.

Finally, we would like to comment on what we believe was achieved by putting the forecasting programme to a group of outside researchers, rather than pursuing the matter internally at the EPO. Advantages included a more panoramic view of the possible techniques that can be applied, a deepening of knowledge and a variety of new techniques to try out. Disadvantages included the time and expenditure that were required to design the programme and the opportunity cost of not being able to progress very far ourselves while it was running. From this experience, we would not seek to persuade other organisations to set up such an external programme, unless they wish to dedicate considerable resources to a fundamental review. Hopefully, the reports that have been given may be useful to others as they contemplate similar issues.

Nevertheless, we appreciate the contributions to the research programme and salute the authors for their dedication and application. We intend to investigate their techniques further, although we will probably use them to augment the traditional methods rather than to replace them entirely. Many questions have been raised for further study. The programme allowed a symbiotic relationship to be created between business planners and a wider group of researchers. So we look forward to further progress. But we should be humble enough to admit that we will never be able to make perfect forecasts[9]. It may actually be to our advantage that this is the case. To some extent at least the future is a question of fate.

---

[9] A statement that is itself a forecast.

References

Abbott A (2006) Rumblings from the fringe. Nature 439:910-911

Adbih Y, Joutz F (2005) Relating the knowledge production function to total factor productivity: an endogenous growth puzzle. International Monetary Fund Working Paper No WP/05/74

Aldrich JH, Nelson FD (1984) Linear probability, logit, and probit models. Sage, London

Armstrong JS (ed) (2001) Principles of Forecasting. Kluwer, Boston

Arellano M, Bond S (1991) Some tests of specification for panel data: Monte Carlo evidence and an application to employment equations. Review of Economic Studies 58:277–297

Arellano M, Bover O (1995) Another look at the instrumental variable estimation of error-components models. Journal of Econometrics 68:29-51

Bainbridge D (1992) Intellectual property. Pitman, London

Baltagi B (2001) Econometric analysis of panel data, 2nd edn. Wiley, New York

Bertin G, Wyatt S (1988) Multinationals and Industrial Property. Harvester-Wheatsheaf, UK/Humanities Press, US

Blind K. (2004) The economics of standards: theory, evidence, policy. Edward Elgar, Cheltenham

Blind K, Grupp H (2000) Gesamtwirtschaftlicher Nutzen der Normung, Volkswirtschaftlicher Nutzen: Zusammenhang zwischen Normung und technischem Wandel, ihr Einfluss auf die Gesamtwirtschaft und den Außenhandel der Bundesrepublik Deutschland. German Institute for Standardisation (Deutschen Institut für Normung) (ed), Berlin

Blind K, Edler J, Frietsch R, Schmoch U (2003a) Erfindungen kontra Patente. Schwerpunktstudie "Zur technologischen Leistungsfähikeit Deutschlands", Fraunhofer Institute for Systems and Innovation Research, Karlsruhe

Blind K, Edler J, Schmoch U, Andersen B, Howells J, Miles I, Roberts J, Hipp C, Green L, Herstatt C, Evangelista R (2003b) Patents in the service industries. Final report, European Commission (ed), Brussels

Blind K, Edler J, Frietsch R, Schmoch U (2004) The patent upsurge in Germany: the outcome of a multi-motive game induced by large companies. Working paper presented at the 8th Schumpeter Conference in Milano, Fraunhofer Institute for Systems and Innovation Research, Karlsruhe

Blundell, R., Bond, S (1998) Initial conditions and moment restrictions in dynamic panel data models. Journal of Econometrics 87:115–143

Bosworth DL (1984) Foreign patent flows to and from the United Kingdom. Research Policy 13:115–124

Bosworth DL (1980) The transfer of US technology abroad. Research Policy 9:378-388

Box GEP, Jenkins GM (1976) Time series analysis forecasting and control. Holden-Day, San Francisco

Brockwell P, Davis R (1986) Time series: theory and methods. Springer Series in Statistics. Springer, Heidelberg
Brockwell P, Davis R (1996) Introduction to time series and forecasting. Springer Texts in Statistics. Springer, Heidelberg
Clements MP, Hendry, DF (1999) Forecasting non-stationary economic time series. MIT Press, Cambridge. MA
Cochran WG (1977) Sampling methods. Wiley, New York
Cohen W, Nelson R, Walsh J (1996) Appropriability conditions and why firms patent and why they do not in the American manufacturing sector. OECD Working Paper
Conover WJ (1999) Practical nonparametric statistics. Wiley, New York
Crepon B, Duguet E (1997) Estimating the innovative function from patent numbers: GMM on count panel data. Journal of Applied Econometrics 12:243–263
Crepon B, Duguet E, Mairesse J (1998) Research, innovation and productivity: an econometric analysis at the firm level. Economics of Innovation and New Technology 7:115–158
Dangerfield BJ, Morris JS (1992) Top-down or bottom-up: aggregate versus disaggregate extrapolations. International Journal of Forecasting 8:233–241
Davison AC (2003) Statistical models. Cambridge
Dernis H, Khan M (2004) Triadic patent families methodology. OECD Directorate for science, technology and industry working paper 2004/2. Available at http://www.olis.oecd.org/olis/2004doc.nsf/linkto/dsti-doc(2004)2
Diebold FX, Lutz K (2000) Unit root tests are useful for selecting forecasting models. Journal of Business and Economic Statistics 18: 265–273
DTI (2001) (Department of Trade and Industry) The 2001 R&D scoreboard: Company data. DTI, London. Available at http://www.innovation.gov.uk/rd_scoreboard/
Draper NR, Smith H (1981) Applied Regression Analysis. Wiley, New York
Durbin J, Koopman SJ (2001) Time series analysis by state space methods. Oxford
Eaton J, Kortum S (1996) Trade in ideas: patenting and productivity in the OECD. Journal of International Economics 40:251–278
Engle R, Granger CWJ (1987) Cointegration and error-correction: representation, estimation and testing. Econometrica 55:251–276
Ericsson NR, Irons JS (1994) (eds.) Testing exogeneity. Oxford
EPIP (2003) (European Policy for Intellectual Property) First conference in Munich. Available at http://www.inno-tec.bwl.uni-muenchen.de/epip/index.html
EPIP (2005) (European Policy for Intellectual Property) EPIP welcome page. Available at http://www.dauphine.fr/imri/EPIP/welcome.html
Epoline (2006) (European Patent Office epoline website) Available at http://www.epoline.org/portal/public

EPO (2006) (European Patent Office) Applicant panel surveys of intentions for filing patent applications at the European Patent Office and other Offices. Survey reports for annual surveys from 2001 to 2005. Authors: Andersen S, Hingley P (2001, 2002, 2003 surveys), European Patent Office with Roland Berger Market Research (2004, 2005 surveys). Available at http://www.european-patent-office.org/aps/

EPO (1994) (European Patent Office) Utilisation of patent protection in Europe. Eposcript 3, European Patent Office, Munich

EPO (2002) (European Patent Office) European Patent Convention, 11th edn. Available at: http://www.european-patent- office.org/epc/pdf_e.htm

EPO (2003) (European Patent Office) European Union, Community Patent – Common political approach of 3 March 2003. Notice from the European Patent Office. Available at http://www.european-patent-office.org/news/info/2003_04_30_e.htm

EPO, JPO, USPTO (2005) (European Patent Office, Japan Patent Office, United States Patent and Trademark Office) Trilateral Statistical Report, 2004 edn. Available at http://www.trilateral.net/tsr/tsr_2004/

European Commission (2002) Summary report of replies to the questionnaire on the impact of the Community utility model with a view to updating the Green Paper on protection by the utility model in the internal market. Document SEC(2001)1307, Available at http://europa.eu.int/comm/internal_market/indprop/docs/model/utilreport_en.pdf

European Commission (2006) Statistical classification of economic activities in the European Community, Rev 1.1. Available at http://europa.eu.int/comm/eurostat/ramon

Evenson R, Puttnam J (1988) The Yale-Canada patent flow concordance. Economic Growth Centre Working Paper, Yale University, New Haven

Fagerberg J (1988) International Competitiveness. Economic Journal 98:355–374

Favero CA (2001) Applied macroeconometrics. Oxford

Fildes R (1992) The evaluation of extrapolative forecasting methods (with discussion). International Journal of Forecasting 8:81–111

Fraunhofer-ISI (Fraunhofer Institute for Systems and Innovation Research), IWW, NIW, DIW, ZEW (2002) Germany's technological performance 2001. Federal Ministry of Education and Research (ed), Bonn

Funke M (1990) Assessing the forecasting accuracy of monthly vector autoregressive models: the case of five OECD countries. International Journal of Forecasting 6:363–378

Gardner ES Jr, McKenzie E (1988) Model identification in exponential smoothing. Journal of the Operational Research Society 39:863-867

Geroski P, Van Reenen J, Walters C (2002) Innovations, patents and cash flow. In: Kleinknecht A, Mohnen P (ed), Innovation and Firm Performance. Palmgrave Macmillan, Basingstoke, pp 31–55

Gibbons JD and Chakraborti S (2003) Nonparametric statistical inference. Chapman and Hall, London

Ginarte J, Park WG (1997) Determinants of patent rights: a cross-national study. Research Policy 26:283–301

Giudici P (2003) Applied data mining. Wiley, Chichester

Greenhalgh C (1990) Innovation and trade performance in the UK. Economic Journal 100:105–118

Greif S, Potkowik G (1990) Patente and Wirtschaftszweige: Zusammenführung der Internationalen Patentklassifikation and der Systematik der Wirtschaftszweige. Carl Heymann, Cologne

Griliches Z (1990) Patent statistics as economic indicators: a survey. Journal of Economic Literature 28:1661–1707

Grupp H, Jungmittag A, Schmoch U, Legler H (2000) Hochtechnologie 2000: Neudefinition der Hochtechnologie für die Berichterstattung zur technologischen Leistungsfähigkeit Deutschlands. Fraunhofer Institute for Systems and Innovation Research, Karlsruhe

Grupp H, Münt G, Schmoch U. (1996) Assessing different types of patent data for describing high-technology export performance. In: Innovation, Patents and Technological Strategies, OECD (ed.), Paris

Hall B, Jaffe AB, Trajtenberg M (2001) The NBER patent citation data file: lessons, insights and methodological tools. National Bureau of Economic Research Working Paper No 8498

Hall B, Ziedonis R (2001) The patent paradox revisited: an empirical study of patenting in the US semiconductor industry, 1979-1995. Rand Journal of Economics 32:101–128

Harhoff D, Scherer F, Vopel K (2003) Citations, family size, opposition and the value of patent rights. Research Policy 32:1343–1363

Hausman J, Hall B, Griliches Z (1984) Econometric models for count data with an application to the patents R&D relationship. Econometrica 52:909–938

Hendry DF (1986) Using PC-GIVE in econometrics teaching. Oxford Bulletin of Economics and Statistics, 48:87–98

Hendry DF (1993) Econometrics: alchemy or science? Blackwell, Oxford

Hendry DF (1995) Dynamic econometrics. Oxford

Hingley P (1997) The extent to which national research and development expenditures affect first patent filings in contracting states of the European Patent Convention. World Patent Information 19:15–25

Hingley P (2004) Analytic estimator densities for common parameters under misspecified models. In: Hubert M, Pison G, Struyf A, Van Aelst S (eds) Theory and applications of recent robust methods. Statistics for Industry and Technology, Birkhauser, Basel

Hingley P (2006) Applications of a technique for estimator densities which allows for model misspecification. Computational Statistics and Data Analysis, submitted

Hingley P, Nicolas M (1999) Improvements to methods for forecasting patent applications using information on patent families. Unpublished paper presented to 19th International symposium on forecasting, Washington DC, 27–30 June 1999

Hingley P, Nicolas M (2004) Methods for forecasting numbers of patent applications at the European Patent Office. World Patent Information 26:191–204

Hingley P, Park WG (2003) Patent family data and statistics at the European Patent Office. WIPO-OECD workshop on statistics in the patent field. Geneva. Available at www.wipo.int/patent/meetings/2003/statistics_ workshop/en/ presentation/statistics_workshop_hingley.pdf

Hyndman RJ, Koehler AB, Snyder RD, Grose S (2002) A state space framework for automatic forecasting using exponential smoothing methods. International Journal of Forecasting 18:439–454

Insightful (2002) S+FinMetrics. Insightful corporation, Seattle, Washington

Insightful (2001) S-PLUS 6 for Windows user's guide. Insightful Corporation, Seattle, Washington

Janz N, Licht G, Doherr T (2001) Innovation activities and European patenting of German firms: a panel data analysis. Paper presented at the Annual Conference of the European Association of Research in Industrial Economics / Working Paper, Centre for European Economic Research (ZEW), Mannheim

Johansen S (1988) Statistical analysis of cointegration vectors. Journal of Economic Control and Dynamics 12:231–54

Johnson D (2002) The OECD Technology Concordance (OTC) Patents by industry of Manufacturer and sector of use, STI working paper 2002/5, Paris OECD.

Joutz FL, Maxwell WF (2002) Modelling the yields on noninvestment grade bond indices: credit risk and macroeconomic factors. International Review of Financial Analysis, 11:345–374

Jungmittag A, Blind K, Grupp H (1999) Innovation, standardization and the long-term production function: a co-integration approach for Germany. Zeitschrift für Wirtschafts- und Sozialwissenschaften 119:205–222

Kalton G (1983) Introduction to survey sampling. Sage, London

Kish L (1995) Survey sampling. Wiley, New York

Kortum S, Lerner J (1997) Stronger protection or technological revolution: what is behind the recent surge in patenting? National Bureau of Economic Research Working Paper 6204, Cambridge MA

Kortum S, Lerner J.(1998) Does Venture Capital Spur Innovation? National Bureau of Economic Research Working Paper No 6846

Kortum S, Lerner J (1999) What is behind the recent surge in patenting? Research Policy 28:1–22

Lanjouw JO, Schankerman M (1999) The quality of ideas: measuring innovation with multiple indicators. National Bureau of Economic Research Working Paper No 7345

Lanjouw JO, Pakes A, Putnam J (1998) How to count patents and value intellectual property: uses of patent renewal and application data. Journal of Industrial Economics 46:405–433

Lee ES, Forthofer RN, Lorimor RJ (1989) Analyzing complex survey data. Sage, London

Levin R, Klevorick A, Nelson R, Winter S (1987) Appropriating the returns from industrial research and development. Brookings Paper on Economic Activity 3:784–829

Li H, Maddala GS, Trost R, Joutz F (1997) Estimation of short run and long run elasticities of energy demand from panel data using shrinkage estimators. Journal of Business and Economics Statistics, 15:

Liao TF (1994) Interpreting probability models. Logit, probit, and other generalized linear models. Sage, London

Löfstedt R (ed) (2003) Riskworld. Journal of Risk Research 6: Issues 4–6

Long JS (1997) Regression models for categorical and limited dependent variables. Sage, London

Lütkepohl H (1993) Introduction to multiple time series analysis, 2nd edn. Springer, Heidelberg

MacKichan Software (2006) Scientific Notebook. Available at http://www.mackichan.com

Maddala GS (2001) Econometrics, Wiley, Chichester

Maskus KE (2000) Intellectual property rights in the global economy. Institute for International Economics, Washington, DC

Meade N (2000) Evidence for the selection of forecasting methods. Journal of Forecasting 19:515–535

Microsoft (2006) Office Online: Excel. http://office.microsoft.com

Miller JR (1998) Spatial aggregation and regional economic forecasting. The Annals of Regional Science 32:253–266

Mills TC (1999) The econometric modelling of financial time series. Cambridge University Press

Nelson CR, Plosser CI (1982) Trends and random walks in macroeconomic time series: some evidence and implications. Journal of Monetary Economics, 10:139–162

Nickell S (1981) Biases in dynamic models with fixed effects. Econometrica, 49:1417–1426

Niiler E (2003) San Diego. Nature 426:690–694

OECD (2003) (Organization for Economic Cooperation and Development) Main science and technology indicators. OECD, Paris. Available from www.oecd.org

OECD (2005a) (Organization for Economic Cooperation and Development) Compendium of patent statistics 2005. Available at http://www.oecd.org/dataoecd/60/24/8208325.pdf

OECD (2005b) (Organisation for economic cooperation and development) Patent statistics task force. 2005 annual meeting summary. Available at http://www.oecd.org/dataoecd/48/24/33938582.pdf

OECD (2005c) See OECD (2003)

OFR (2004a) (The Office of the Federal Register) Federal Register 69(31). National Archives and Records Administration, p 7454

OFR (2004b) (The Office of the Federal Register) Federal Register 69(149). National Archives and Records Administration, p 47123

Pakes A (1986) Patents as options: some estimates of the value of holding European patent stocks. Econometrica 54:755–784

Pampel FC (2000) Logistic regression: a primer. Sage, London

Pankratz A (1983) Forecasting with univariate Box-Jenkins models: concepts and cases. Wiley, New York

Park WG (2001) International patenting, patent rights, and technology gaps. Working Paper, Department of Economics, American University

Pastor L (2002) A model weighting game in estimating expected returns. In: Mastering investment, your single source guide to becoming a master of investment. J Pickford (ed), FT Prentice-Hall, Harlow, UK

Preez J du, Witt F (2003) Univariate versus multivariate time series forecasting: an application to international tourism demand. International Journal of Forecasting 19:435–451

Putnam J (1996) The value of international patent rights. PhD thesis, Yale University, New Haven, CT

Ramsay JO, Silverman BW (1997) Functional data analysis. Springer, Heidelberg

Rossana RJ, Seater JJ (1995) Temporal aggregation and economic time series. Journal of Business and Economic Statistics 13:441–451

Rutterford J (1983) Introduction to stock exchange investment. Macmillan, London

Särndel CE, Swensson B, Wretman J (1992) Model assisted survey sampling. Springer, New York

SAS Inc (2006) Analytics. Available at http://www.sas.com/technologies/analytics/index.html

Schankerman M, Pakes A (1986) Estimates of the value of patent rights in European countries during the post-1950 period. Economic Journal 97:1–25

Schiffel D, Kitti C (1978) Rates of invention: international patent comparisons. Research Policy 7:324-340

Schmoch U, Laville F, Patel P, Frietsch R (2003) Linking technology areas to industrial sectors. Report to the European Commission, DG Research, Fraunhofer Institute for Systems and Innovation Research, Karlsruhe, Paris, Brighton

Schmoch U, Koschatzky K (1996) Freie Erfindungen erfolgreich verwerten. In: Grupp H (ed) Schriftenreihe Zukunft der Technik. TÜV Rheinland, Cologne

Schwartzkopf AB, Tersine RJ, Morris JS (1998) Top-down versus bottom-up forecasting strategies. International Journal of Production Research 26:1833–1843

Shumway R, Stoffer D (1999) Time series analysis and its applications. Springer Texts in Statistics, Springer, Heidelberg

Slama J (1981) Analysis by means of a gravitation model of international flows of patent applications in the period 1967-1978. World Patent Information 3:1–8

Stock JH (1988) Estimating continuous-time processes subject to time deformation. Journal of the American Statistical Association 83:77–85

Swamy PAVB, Tavlas GS (2005) Theoretical conditions under which monetary policies are effective and practical obstacles to their verification. Economic Theory 25:999–1005

Tong H (1990) Nonlinear time series: a dynamical systems approach. Oxford University Press

UN (2006) (United Nations statistics division) ISIC Rev. 3.1, detailed structure and explanatory notes. Available at http://unstats.un.org/unsd/cr/registry

UNESCO (1980–2002) (United Nations Educational, Scientific and Cultural Organisation) Statistical Yearbook (1980–2002). Paris

Verspagen B, Morgastel T v, Slabbers M (1994) MERIT concordance table: IPC-ISIC (rev. 2). MERIT Research Memorandum, Maastricht

Wakelin K (1997) Trade and innovation: theory and evidence. Edward Elgar, Cheltenham

Wei W (1990) Time series analysis. Addison-Wesley, New Jersey

Wilcox R (2005) Robust estimation and hypothesis testing. Elsevier, Burlington, MA

Woolridge JM (2002) Econometric analysis of cross section and panel data. MIT Press, Cambridge, MA

World Bank (2002) World development indicators. Washington, DC

WIPO (2000) (World Intellectual Property Organisation) International Patent Classification, guide, survey of classes and summary of main groups. 7th edn, vol 9. Geneva

WIPO (2003) (World Intellectual Property Organisation) Workshop on statistics in the patent field, session on Patents by industry. http://www.wipo.int/patent/meetings/2003/statistics_workshop/en/program.htm

WIPO (2004) (World Intellectual Property Office) WIPO-OECD Workshop on the use of patent statistics, with the support of EPO, JPO and USPTO. Available at http://www.wipo.int/patent/meetings/2004/statistics_workshop/en/

WIPO (2005) (World Intellectual Property Organisation) International Patent Classification. Available at: http://www.wipo.int/classifications/ipc/en/preface.htm

WIPO (2006a) (World Intellectual Property Organization) Paris Convention for the Protection of Industrial Property, of March 20, 1883, as amended September 28, 1979. Available at http://www.wipo.int/treaties/en/ip/paris/pdf/trtdocs_wo020.pdf

WIPO (2006b) (World Intellectual Property Organisation) PCT Newsletter. Available at http://www.wipo.int/pct/en/newslett

Zellner A (1962) An efficient method of estimating seemingly unrelated regressions and tests for aggregation bias. Journal of the American Statistical Association 57: 348–368

Zhao JX, Foulkes AS, George EI (2005) Exploratory Bayesian model selection for serial genetics data. Biometrics 61:591–599

Zhou H (2004) Forecasting PCT filings. Presented at WIPO-OECD Workshop on the use of patent statistics, with the support of EPO, JPO and USPTO. Available at http://www.wipo.int/patent/meetings/2004/statistics_workshop/en/presentations/statistics_workshop_zhou.pdf

Zivot E, Wang J (2003) Modeling financial time series with S-PLUS. Insightful Corporation, Seattle

Name index

Abbott, A., 242
Abdih, Y., 27, 38
Aldrich, J.H., 171
Andersen, B., 168
Andersen, S., 194, 204, 207, 213, 222, 223
Arellano, M., 37, 133
Armstrong, J.S., 48, 201

Bainbridge, D., 2
Baltagi, B., 157
Bertin, G., 219
Blind, K., 73, 74, 81, 86, 162, 168, 184, 208
Blundell, R., 37
Bond, S., 37, 133
Bosworth, D.L., 129
Bover, O., 37
Box, G.E.P., 49, 100, 211
Brockwell, P., 98

Chakraborti, S., 235
Clements, M.P., 36
Cochran, W.G., 223
Cohen, W., 128
Conover, W.J., 63
Crepon, B., 219

Dangerfield, B.J., 48
Davis, R., 98
Davison, A.C., 233
Dernis, H., 6, 217

Diebold, F.X., 36
Dikta, G., 95, 211
Doherr, T., 73, 74, 184
Draper, N.R., 192, 196
Duguet, E., 219
Durbin, J., 206

Eaton, J., 27, 129
Edler, J., 81, 86, 162, 168, 184
Engle, R., 36
Ericsson, N.R., 36
Evangelista, R., 168
Evenson, R., 75, 76

Fagerberg, J., 73
Favero, C.A., 38
Fildes, R., 53
Forthofer, R.N., 162, 222
Foulkes, A.S., 233
Frietsch, R., 75, 76, 81, 86, 159, 162, 184, 219
Funke, M., 49

Gardner, E.S. Jr., 51
George, E.I., 233
Geroski, P., 184
Gibbons, J.D., 235
Ginarte, J., 32
Giudici, P., 244
Granger, C.W.J., 36
Green, L., 168
Greenhalgh, C., 73

Greif, S., 76
Griliches, Z., 187, 219
Grose, S., 51, 206
Grupp, H., 73, 74, 89, 168, 172

Hall, B., 130, 187
Harhoff, D., 9, 151, 203
Hausman, J., 38, 187
Hendry, D.F., 34, 36, 192
Herstatt, C., 168
Hingley, P., 1, 6, 27, 46, 85, 86, 97, 109, 131, 191, 193, 194, 204, 207, 213, 217, 222, 223, 233, 235
Hipp, C., 168
Howells, J., 168
Hubert, M., 235
Hyndman, R.J., 51, 206

Irons, J.S., 36

Janz, N., 73, 74, 184
Jenkins, G.M., 49, 100, 211
Johansen, S., 35
Johnson, D., 76
Joutz, F., 27, 36, 37, 38
Jungmittag, A., 73, 89

Kalton, G., 162, 164
Khan, M., 6, 217
Kish, L., 163, 164
Kitti, C., 128
Kleinknecht, A., 184
Klevorik, A., 128
Koehler, A.B., 51, 206
Koopman, S.J., 206
Kortum, S., 27, 73, 74, 129, 184
Koschatzky, K., 168

Lanjouw, J.O., 151
Laville, F., 75, 76, 81
Lee, E.S., 162
Lee, S., 222
Legler, H., 89
Lerner, J., 27, 73, 74, 129, 184
Levin, R., 128

Li, H., 37
Liao, T.F., 171
Licht, G., 73, 74, 184
Löfstedt, R., 241
Long, J.S., 171, 175
Lorimor, R.J., 162, 222
Lütkepohl, H., 98, 104, 106
Lutz, K., 36

Maddala, G.S., 37, 236
Mairesse, J., 219
Maxwell, W.F., 36
McKenzie, E., 51
Meade, N., 41, 51, 205
Miles, I., 168
Miller, J.R., 48, 71
Mills, T.C., 50
Mohnen, B., 184
Morgastel, T., 75
Morris, J.S., 48
Münt, G., 172

Nelson, C.R., 36
Nelson, F.D., 171
Nelson, R., 128
Nickell, S., 37
Nicolas, M., 1, 191, 193, 217
Niiler, E., 7

Pakes, A., 151
Pampel, F.C., 171
Pankratz, A., 49
Park, W.G., 6, 27, 32, 46, 97, 125, 129, 131, 204, 215
Pastor, L., 221
Patel, P., 75, 76, 81
Pickford, J., 221
Pison, G., 235
Plosser, C.I., 36
Potkowik, G., 76
Preez, J du., 48
Putnam, J., 29, 75, 76, 151

Ramsay, J.O., 235
Roberts, J., 168

Rossana, R.J., 49, 71
Rutterford, J., 221

Särndel, C.E., 223
Schankerman, M., 151
Scherer, F., 151
Schiffel, D., 128
Schmoch, U., 75, 76, 81, 86, 89, 162, 168, 172, 184
Schwartzkopf, A.B., 48
Seater, J.J., 49, 71
Shumway, R., 98
Silverman, B.W., 235
Slabbers, M., 75
Slama, J., 129
Smith, H., 192, 196
Snyder, R.D., 51, 206
Stock, J.H., 49
Stoffer, D., 98
Struyf, A., 235
Swamy, P.A.V.B., 235
Swensson, B., 223

Tavlas, G.S., 235
Tersine, R., 48
Tong, H., 33

Trost, R., 27, 37

Van Aelst, S., 235
Van Reenen, J., 184
Verspagen, B., 75
Vopel, K., 151

Wakelin, K., 73
Walsh, J., 128
Walters, C., 184
Wang, J., 104, 105, 106
Wei, W., 98
Wilcox, R., 235
Winter, S., 128
Witt, S.F., 48
Woolridge, J.M., 139
Wretman, J., 223
Wyatt, S., 219

Zellner, A., 236
Zhao, J.X., 233
Zhou, H., 221
Ziedonis, R., 130
Zivot, E., 104, 105, 106

Index

administrative council (AC), 198
aerospace and defence, 181
agent fees, 30
aggregation, 48, 79, 126, 136, 214
aircraft, 91
Akaike's information criterion (AIC), 106, 114
analytical business enterprise research and development (ANBERD), 20, 81
annual turnover, 167
appeal, 199
applicant, 2
applicant name, 179, 220
applicant panel survey, 14, 160, 170, 194, 199, 204, 220, 232
ARMAX, 101
Asian crisis, 172
attorney, 235
Australia, 149, 150, 164
Austria, 167
autocorrelation, 113, 115, 139, 152
autocorrelation function (ACF), 98, 111, 115
autocovariance function, 98, 99, 101
autoregression, 100, 133, 139, 192, 221
autoregressive distributed lag (ADL), 102, 106, 111, 211, 213, 225

autoregressive integrated moving average (ARIMA), 12, 21, 23, 33, 41, 48, 55, 100, 205, 211, 226
autoregressive moving average model (ARMA), 100, 102, 105, 211

backward shift operator, 100
base metals, 91
basic assumptions (BA) document, 198
Bayesian Information Criterion (BIC), 50, 59
Bayesian model averaging (BMA), 233
Bayesian model discrimination (BMD), 233, 235
Belgium, 85, 86, 226
BERD financed by industry, 225, 230
best paper award, 25
biggest group, 195
bilateral patent families, 7
biotechnology, 128, 143
biotechnology/dot-com bubble, 1
bloc, 41
blocking, 130
bottom up, 48
bounded influence estimators, 18, 204

Box-Cox transformation, 99, 212, 231
Brazil, 149
breadth of patenting, 126
budget and finance committee (BFC), 198
Bulgaria, 150
business activity, 241
business cycle, 221
business enterprise research and development (BERD), 97, 98, 135, 212, 215
business planning, 201, 220, 235
Business Week, 20

Canada, 20, 75, 76, 150, 152, 164
Canadian SIC, 75
capital asset pricing model, 221
capital investment, 178
capital stock, 35
changing political criteria, 242
chaos theory, 15
chemicals, 77, 89, 143, 181
chief economist, 204
China, 149, 150
claim, 149
classification, 8, 75
client database system, 220
cointegration, 34, 36
collapse of the new economy, 96
communications, 169, 173, 188
Community innovation survey (CIS), 12, 160
companion matrix, 104
company size, 74
comparison of forecasts, 232
competition, 82
competitive effect, 240
computers, 96, 143, 147, 206
concomitant variable, 33
concordance, 74, 75, 76, 81, 208
concordance matrix, 21, 80, 209
conditional forecast, 105, 116, 120
conference, 25, 26, 204
confidence interval, 22, 192
confidence limits, 200, 216, 223

conformable operator, 36
consolidation, 43
construction, 206
consumption expenditure, 35
continuity, 219
continuous applicant, 171, 173, 187
copyright, 182
correlation, 71, 80, 93, 139, 184, 219, 224
cost of funds, 178
costs of patenting, 218
count data, 34, 38
count model, 175, 187
country, 176
country specific transfer matrix, 80
Court of Appeals of the Federal Circuit, 130
covariance analysis, 221
Creditreform, 167
critical t value, 234
cross correlation, 109, 212, 221
cross correlation function (CCF), 101, 115
cross covariance function, 102
cross section, 49, 71, 129, 219
cubic trend, 106

decision theory, 27, 129
deflator, 85
degrees of freedom, 59, 171
Delphi technique, 240
Denmark, 86
Department of Trade and Industry (DTI) scoreboard, 161, 177, 178, 186, 222
designation fees, 205
Deutsche Bundesbank, 25
developing countries, 155, 217
differencing, 39, 99, 107, 115, 158, 209, 214, 231
diffusion, 125, 129, 150, 152
disaggregation, 42, 75, 125, 205, 216
discontinuous applicant, 171, 187
disposable income, 35
disruption of trend, 192

Index 261

distribution of applicants, 162
DOCDB (EPO patent publications database), 97
domestic filing, 34, 149, 193, 237
Dun & Bradstreet (D&B), 78, 167
dynamic linear model (DLM), 41, 50, 60, 205
dynamic multivariate regression, 102
dynamic panel data technique, 37
dynamic specification models, 130

econometric modelling, 13
economic cycle, 232
elasticity, 28, 139, 144, 152, 219
electrical equipment, 77, 89
electricity, 56, 144
electricity, gas, oil or steel, 181
electronics, 167, 181
emerging areas of technology, 8
emerging markets, 155
endogenous effectors, 38
energy, 173, 188
EPO budget, 198, 231
EPO experts, 17, 22
equilibrium relationship, 35
error correction, 34
error correction model (ECM), 36, 192, 215
estimation bias, 37
euro-direct (EPC filing), 4
euro-PCT-IP (PCT international phase filing), 4
euro-PCT-RP (PCT regional phase filing), 5
European Commission, 168, 203, 220
European Patent Convention (EPC), 4
European Patent Office (EPO), 1
European Policy for Intellectual Property (EPIP), 203
Eurostat, 77
examiner, 235
exchange rate, 35
exogeneity, 36, 85

exponential smoothing, 12, 206, 235
export flows, 74
exports, 82, 87, 91, 209
externality, 20
extrapolation, 48

feedback (from patents to R&D), 156
fees, 205, 218
fertile technology hypothesis, 129
filing costs, 31, 139
filing fees, 30
filings flows, 14
filings per worker, 138
final action, 199
finance activity, 241
finance department, 199
finite population correction, 163, 223
Finland, 75, 86, 87, 94
firm, 7, 14, 17, 19, 20, 73, 127, 219
firm productivity, 29
first filing, 2, 11, 22, 34, 136, 148, 192, 197, 201, 213, 236
first national filings, 97
first-to-invent rule, 194
flow, 5, 15, 21, 34, 125, 194, 218, 236, 240
food, 91
forecast bias, 235
forecast error variance decomposition, 106
forecast performance rating procedure, 202
forecasting competition, 235
foreign trade, 73, 77
fractal structure, 8
France, 19, 81, 86, 97, 164, 165, 171, 180, 226
free inventor, 160, 173, 209
Friedman test, 62
functional data analysis, 235

Gaussian white noise (GWN), 100, 109, 113, 116
GDP per capita, 139

generalized least squares (GLS), 133, 139, 147, 152, 157
generalized method of moments (GMM), 37, 141, 147, 152, 158
general-to-specific econometric modelling approach, 34
geometric mean, 70
Germany, 7, 19, 74, 76, 81, 86, 87, 97, 149, 154, 162, 164, 165, 171, 180, 209, 217, 226
global model, 236
globalisation, 205, 218
globalisation of consumption, 236
globalisation of production, 236
GMM-type estimators, 37
government, 135
government-financed BERD, 231
grace period, 202
Granger-causality, 104
grant, 160
grant publication, 199
gravity model, 129
Greece, 149
gross domestic expenditure on research and development (GERD), 135, 215
gross domestic product (GDP), 7, 32, 34, 38, 98, 111, 115, 118, 203, 209, 213
gross national product (GNP), 129
group of experts on forecasting, 9
Gulf war, 96, 205

handling and processing, 143
Herfindahl-index, 82
heteroscedasticity, 94, 99, 210
hierarchical approach, 215
high-technology, 89
holding company, 178
holdings, 169
Hoppenstedt, 167
horizon, 54, 66, 195, 205, 232
human necessities, 143
Hungary, 137, 150

identifiability, 33

identification, 102, 236
imitation, 29, 127
impulse response analysis, 106, 120
INCENTIM, 160
industrial applicability, 4
industrial standard, 209
industrial structure, 80
industry, 7, 20, 41, 81, 135
information and communication technologies (ICT), 172
infringement, 127
innovation, 105, 119, 127, 130, 134, 150, 157
innovation process, 99
inorganic chemistry, 206
institutional factor, 134
intangible asset, 127
interest rate differential, 35
internal performance, 128
international firm databases, 167
international patent classification (IPC), 8, 20, 43, 75, 81, 198, 206, 208
internet, 18, 204
internet search, 168, 179
intervention analysis, 38
invention, 28, 109, 236
inventive step, 29
inventiveness, 4
inventor, 2, 194, 235
inventory, 35
Ireland, 149
ISIC, 8, 75, 77, 173, 198, 208
Israel, 150
IT hardware production and distribution, 181
Italy, 86, 97, 164, 165, 171

Japan, 19, 41, 42, 55, 81, 86, 87, 97, 137, 148, 149, 154, 164, 171, 180, 188, 193, 207, 212, 217, 226
Japan Patent Office (JPO), 7, 201, 204, 218, 238
Johnson concordance, 76
joint cluster, 5, 131, 156, 191, 195, 197, 206, 216, 223, 235

JPO survey, 160

knowledge capital, 155
Korea, 149, 150
KPSS test, 105, 107

labour, 225
lag, 47, 59, 92, 100, 109, 134, 156, 178, 186, 209, 210, 212, 225
large companies, 223
lead time, 30, 128
least squares regression, 158
leather, 91
licensing, 128
likelihood function, 234
linear model, 191, 212
linear regression, 33, 161, 175, 212, 215
linear trend, 106, 202, 230
Ljung-Box test, 105, 109, 113, 114
logistic model, 192
logistic regression, 161, 187, 219
long memory model, 113
long run interest rate, 35
long term planning, 240
low-technology, 89, 210

machinery, 77, 89
main science and technology indicators (MSTI), 8, 212, 225
manpower capacity, 199
manpower planning, 42
manufacturing, 89, 143, 167, 169, 188, 209
market, 3, 28, 134, 151, 155, 236
market capitalisation, 177, 185, 186
Market Europe, 167
matrix, 33
maximum likelihood, 50
mean absolute percentage error (MAPE), 53, 55, 66, 232
mean function, 98
media, 181
medical technology, 89
medicine, 231

medium term business plan (MTBP), 199
merger and acquisition, 178
MERIT Concordance, 75
Mexico, 150
micro approach, 159
micro econometric model, 19
Microsoft Excel, 161, 225
Microsoft Word, 161
missing observations, 135
mode of filing, 131, 225
module (of research programme), 15, 203
Monaco, 149
monolateral patent families, 7
monthly data, 47
monthly forecasting, 205
motor vehicle, 89
moving average process, 100
multidimensional imputation model, 180
multiplicative model, 50
multivariate Gaussian white noise, 103
multivariate normal distribution, 103

NACE, 8, 20, 21, 75, 81, 173, 198, 208
nation, 125
National Bureau of Economic Research (NBER) patent citation database, 160
national filing, 238
national first filing, 109, 169
national level, 20
national office, 5, 11, 179, 193, 218
national patent procedures, 2
National Science Foundation, 203
negative binomial regression, 161, 187
negative exponential distribution, 29
Netherlands, 81, 86, 87, 97
neural network, 15
new knowledge, 35
New Zealand, 150, 164
nonlinear time series regression, 33

non-metallic products, 91
non-parametric methods, 235
non-response, 169, 224
Norway, 150
novelty, 4
now-casting, 217
number of employees, 19, 167, 169, 177, 184, 219
number of researchers, 211

Observatoire des Sciences et des Techniques (OST), 78, 160
office machinery, 89, 91
oil, 91
opposition, 199
optimum allocation, 163
option, 215
oracle at Delphi, 96
Organisation for Economic Co-operation and Development (OECD), 7, 20, 77, 98, 202, 212, 220, 225
orthogonal forecast error variance decomposition (EVD), 104
orthogonal impulse response, 104
orthogonal shock, 119
other transport, 89, 91
over dispersion, 175

panel data, 125, 135
paper, 91
Paris Convention, 3, 127, 191, 192, 238
parsimony, 2, 52
partial adjustment model, 36
Parzen window, 218
Patent Cooperation Treaty (PCT), 2, 4, 22, 126, 169, 172, 176, 187, 194, 199, 206, 214, 221, 238
patent family, 6, 97, 131, 148, 150, 156, 198, 216, 217, 236
patent law, 126
patent portfolio, 130
patent quality, 28
patent renewal, 151
patent statistics task force, 202, 220

patent value, 30, 151, 215
PCT designation, 238
pensum, 206, 211
Perinorm, 81
petroleum, 89, 91
pharmaceuticals, 89, 206
pharmaceuticals, health or personal care, 181
Philippines, 150
physics, 56, 143
poisson regression, 161
Poland, 150
policy implication, 38
political change, 241
population, 129, 165
Portugal, 149
posterior weight, 233
post-stratification, 164, 223
pre-stratification, 164
prewhitening, 109, 212, 231
priority, 6, 11, 13, 34, 127, 134, 150, 161, 163, 171, 219, 238
priority number, 148
product type, 217
production, 73
productivity, 127
profit, 29, 177, 186
profit flow, 31
pro-patent hypothesis, 129
proportional allocation, 163
proportional sample, 163
proprietary right, 134
provisional filing, 194
public administration, 169
public research, 169, 173, 188
publication, 2, 74, 198
purchasing power parity (PPP), 225
pure and applied organic chemistry, 143

quadratic trend, 106
quantile regression, 161
quasi-likelihood ratio (QLR) test, 139
questionnaire, 12, 17, 194, 204, 223
R&D efficiency, 130

R&D employees, 74, 167
R&D expenditure, 4, 7, 31, 72, 77, 133, 167, 177, 180, 183
R&D stock, 28, 218
R^2, 83, 240
radio and television, 89
random effects estimation, 152
random group, 195
random sample, 11, 161, 219, 222
recurrent filings, 220
regional office, 28
regional procedure, 2
regression, 7, 12, 33, 127, 138, 187, 191, 203, 211, 221, 232, 237, 240
reputation, 30, 128
research and development (R&D), 13, 19, 28, 34, 93, 98, 109, 111, 114, 118, 127, 141, 147, 150, 156, 157, 160, 180, 182, 183, 186, 195, 203, 209, 218, 220, 224, 230, 240
research competition, 15, 25, 26, 204
researcher in residence, 204
response rate, 18
risk assessment, 241
risk taking, 157
robust regression methods, 235
robustness, 19, 235
Romania, 137, 150
root mean square error (RMSE), 53, 66, 232
root mean square proportion error (RMSPE), 133, 141, 147, 152, 216, 232
round table meeting, 198
Russia, 149, 150

S+FinMetrics, 105
sales, 19, 35, 128, 167, 168, 170, 177, 180, 219
sample, 17
sample of applicants, 188
sampling fraction, 163
SAS, 225, 230
saturating model, 207
Scandinavia, 180

scenario building, 240
scenario forecast, 120, 232
science and engineering personnel, 32
Scientific Notebook, 238
search, 4, 199
seasonal factors, 51
seasonal modelling, 71
secrecy, 30, 128
sector, 74, 81, 87, 88, 126, 167, 173, 209, 219
sector size, 176
seemingly unrelated regression, 236
seemingly unrelated time series equations (SUTSE) model, 52
self determining approach, 33
semiconductors, 130
September 11th 2001, 96, 172
services, 128, 167, 168
Shapiro-Wilks test, 105, 109
short run interest rate, 35
Siemens, 167, 223
sigmoidal model, 207
simultaneous equations, 37, 236, 238
simultaneous forecast, 120
skewed distribution, 175
skewness, 162, 224
small and medium sized enterprises (SME), 160, 169, 224
small economies, 172
software, 128
software, IT services, telecommunication and support, 181
South Africa, 150, 164
Spain, 85, 94, 149
special search, 199
spillover, 135
S-PLUS, 105, 107, 225
SPRU, 160
SPSS, 161, 175
stability test, 119
STATA, 157, 161, 175, 225
state space, 206

state space model, 33, 48, 51, 213, 231
stationarity, 35, 47, 99, 102, 103, 105, 106, 115, 212
stationary transformation, 49
stepwise reduction technique (SRT), 106, 113
stock, 28, 74
Stock Exchange Commission (SEC), 19
stock market bubble, 205
stock of knowledge, 35
stock of standards, 82, 83, 86, 90
straight line regression, 1, 192
strategic patenting, 86
stratification, 163
stratified sample, 162, 167, 222
strength of patent rights, 29, 31
structural model, 14, 237
Structural Time series Analyser, Modeller and Predictor (STAMP), 49, 51
subsequent filing, 3, 11, 34, 109, 136, 148, 156, 192, 198, 213, 217, 236, 238
substantive examination, 4, 199
supranational office, 11
supranational patent procedure, 2
survey, 11, 14, 17, 127, 160, 204
Sweden, 85, 86, 87
Switzerland, 149, 167
system of equations, 35

technical classification, 217
technical field, 205
technical rule, 74
technical services, 169
technical standard, 74, 81
technological field, 131, 136, 143
technological knowledge, 90
technological output, 73
technological progress, 34
technology, 75, 81, 131, 225
technology group, 143, 147, 216
technology trend watching, 8
telecommunications, 143, 147

terrorism, 96
textile, 91
time series, 23
Time Series Processor (TSP), 49
timeliness, 148, 151, 154, 156, 203, 243
tobacco, 91
top down, 48, 168
tourism, 48, 205
trade, 127, 129
trademark, 182
transfer coefficient, 13, 22
transfer effect, 5, 218
transfer function, 206, 212, 226
transfer function model (TFM), 101
transfer method, 13, 21, 191, 192, 197, 207, 234, 236, 237
transfer ratio, 156, 193
transformation, 59
transition matrix, 20, 104, 220
translation fees, 30
trend analysis, 191, 196
trend homogeneity, 60
trend method, 234
trend variable, 209
triadic patent families, 7
trilateral global applicant panel survey, 12
trilateral office, 237
trilateral patent family, 7, 148, 150, 152, 153, 154
trilateral statistical working group, 202, 236
Turkey, 164
turning point, 232

UK, 19, 81, 86, 87, 97, 129, 149, 164, 171, 180, 226
unemployment, 48
unit root, 35
United Nations Educational, Scientific and Cultural Organisation (UNESCO), 135
United States Patent and Trademark Office (USPTO), 7, 75, 182, 201, 202, 238

university, 135, 160
US, 19, 41, 55, 81, 86, 87, 97, 128, 129, 137, 148, 154, 164, 171, 172, 180, 188, 193, 206, 207, 212, 217, 226
USPC, 75
USPTO survey, 160
US-SIC, 76

value added, 19, 77, 82, 86, 209
vector, 33
vector autoregression (VAR), 13, 23, 33, 49, 50, 58, 102, 105, 118, 205, 211, 214

wearing and dressing, 91
weight, 165, 226
weighted least squares, 210

welfare, 127
what-if analysis, 232
white noise, 99, 103
Winsorised estimation, 204
wood, 91
workforce size, 215
workload, 199, 202
World Bank, 135
World Base, 167
World Intellectual Property Organisation (WIPO), 4, 126, 202, 204
world wide web, 219, 231

Yale-Canada patent flow concordance, 75
Year-2000 computer problem, 96

Printing: Krips bv, Meppel
Binding: Stürtz, Würzburg